Quantum Shift
in the
Global Brain

How the New Scientific Reality Can Change
Us and Our World

Ervin Laszlo

Inner Traditions
Rochester, Vermont

Inner Traditions
One Park Street
Rochester, Vermont 05767
www.InnerTraditions.com

Library of Congress Cataloging-in-Publication Data
Laszlo, Ervin, 1932–
  Quantum shift in the global brain : how the new scientific reality can change
us and our world / Ervin Laszlo.
      p. cm.
  Includes bibliographical references and index.
  Summary: "The shift from scientific materialism to a multidimensional
worldview in harmony with the world's great spiritual traditions"—Provided
by publisher.
  ISBN-13: 978-1-59477-233-7
  ISBN-10: 1-59477-233-9
  1. Civilization, Modern—1950– 2. Science and civilization. 3. Science—
Social aspects. 4. Human evolution. 5. Social change. 6. Social evolution. 7.
Consciousness—Social aspects. 8. Spirituality—Social aspects. 9. Paradigms
(Social sciences) 10. Club of Budapest. I. Title.
  CB430.L37 2008
  909.82'5—dc22
                                                                    2007044283

Printed and bound in the United States by Lake Book Manufacturing

10   9   8   7   6   5   4   3   2   1

Text design and layout by Jon Desautels
This book was typeset in Sabon with Caslon Pro used as a display typeface

To send correspondence to the author of this book, mail a first-class letter
to the author c/o Inner Traditions • Bear & Company, One Park Street,
Rochester, VT 05767, and we will forward the communication.

dedicated to

Jenna,

Ishana,

Kahlia Paola

—and all the young people

who will lead the quantum shift to a better world

Contents

PART THREE

GLOBAL*SHIFT* IN ACTION: THE CLUB OF
BUDAPEST AND ITS INITIATIVES

What is a quantum shift in the global brain?

The global brain *is the quasi-neural energy- and information-processing network created by six and a half billion humans on the planet, interacting in many ways, private as well as public, and on many levels, local as well as global.*

A quantum shift *in the global brain is a sudden and fundamental transformation in the relations of a significant segment of the six and a half billion humans to each other and to nature—a* macroshift in society—*and a likewise sudden and fundamental transformation in cutting-edge perceptions regarding the nature of reality—a* paradigm shift in science. *The two shifts together make for a veritable "reality revolution" in society as well as in science.*

The Reality Revolution

In the first decade of the twenty-first century we face a new reality, individually as well as collectively. Our reality is shifting because the human world has become unstable and is no longer sustainable. But the reality revolution harbors a unique opportunity. This decade is the first in history that offers the choice between being the last decade of a fading, obsolete world or the first of a new and viable one.

The emerging reality is radically new. We are experiencing ever more frequent and ever greater shocks and surprises, and these are not due simply to blindness and ignorance. It is our reality that is shifting. As the economist Kenneth Boulding remarked, the only thing we should not be surprised at is being surprised.

The new reality is an intrinsically surprising reality. Nothing continues in the same way as it did before; everything "bifurcates." This expression, coming originally from mathematics and chaos theory, indicates that the path of development of a system encounters a rapid, previously unforeseen change. We live in an age of bifurcation in the midst of a fundamental transformation of our world: in a *Macroshift*.

The reality shift we experience regards the way we relate to each other, to nature, and to the cosmos. Previously some of us had suspected that this reality might soon shift, but the great bulk of humanity proceeded on the assumption that things would remain pretty much the same as they were: business as usual. But in the year 2007, it is daily becoming more evident that business is definitely not as usual. The Earth is literally transforming under our feet. On New Year's eve the Russians celebrated in the former Red Square without a trace of ice and snow; in January New Yorkers walked in Central Park in shirtsleeves; the center of Greenland is taken up by an unfrozen lake the size of Lake Michigan, Lake Superior, and Lake Erie combined; and there is hardly any of the legendary snow left on top of Kilimanjaro. Anyone who still doubts that the world we live in is changing must be blind, obstinate, or just plain stupid.

Of course, the climate is just one of many changes under way, though it is the most visible. Connected with climate change are a host of other factors that are just as prone to change as the ecology, in the economic, social, political, and cultural arenas. The bottom line is that, in more respects than one, proceeding further as we have up till now takes us to a catastrophic bifurcation: to a fateful tipping point.

Change is no longer mere theory, and it is no longer merely an option: it is a reality, an imperative of our survival. Proceeding on the assumption of business as usual (BAU) is suicidal.

Interestingly and importantly, our map of the world is also changing: science itself is in the midst of a paradigm shift. The new paradigm gives us a deeper understanding of the nature of quantum shifts in complex systems in nature as well as in society. Complex systems do not evolve smoothly, step by step: they are highly nonlinear. They evolve step by step only up to a point, then they reach a threshold of stability and either break down or bifurcate. This is true of the evolution of stars (at a given point they either explode as a supernova and spew forth the matter that will become the stuff of the next generation of stars or they collapse into a black hole); it is true of living species (sooner or later in their lifespan most species are threatened with extinction—and then they either mutate into a more viable species or become extinct); and it is also true of entire civilizations (they too evolve or go under, as the experience of the communist world demonstrated in the winter of 1989/90).

Does this mean that human society may be doomed, and we may become extinct even as a species? The currently dominant form of civilization does seem to have reached its limits and is bound to change. But our demise as a species, while it cannot be excluded, is by no means decided. We have enormous and as yet unexploited resources for coping with the challenges that face us. We have a whole range of new and sophisticated technologies at our disposal, and radically new insights are emerging at the cutting edge of the sciences.

However, the key insight coming from the new paradigm in the sciences is not technological. It is the confirmation of something people have always felt but could not give a rational explanation for: our close connection to each other and to the cosmos. Traditional people have known of it and have lived it, but modern civilization has first neglected and then denied it. Yet genuine spiritual experience offers direct evidence of our links to each other and to all of creation, and now science confirms the validity of such intuitions.

Until the last decade or two, scientists and science-minded people considered the feeling of human and human-nature interconnection a mere delusion. Then the evidence started to come in. A fresh look at our connections in the framework of the new sciences—quantum physics above all—began to indicate that the "oneness" people sometimes experience is not delusory and that the explanation of it is not beyond the ken of the sciences. As quanta, and entire atoms and molecules, can be instantly connected across space and time, so living organisms, especially the complex and supersensitive brain and nervous system of evolved organisms, can be instantly connected with other organisms, with nature, and with the cosmos as a whole. This is vitally important, for admitting the intuition of connections to our everyday consciousness can inspire the solidarity we so urgently need to live on this planet—to live in harmony with each other and with nature.

The oracle at Delphi advised, "Know thyself." We should complete this by saying, "Know thyself as part of an interconnected rapidly changing world." As this book will show, this knowledge and the practical wisdom that follows from it have become the precondition of the persistence of human civilization and even the survival of the human species.

THE BIRTH AND BODY OF THIS BOOK

This book embraces, for the first time, both sides of my lifelong interests and research: the practical side, focused on the problems, opportunities, and challenges we now face individually as well as collectively, and the theoretical side, seeking the contours of the reality suggested by the latest developments in the sciences. Together, the two sides provide essential orientation for an epoch of quantum shifts: a time when the terrain is changing under our feet and so is our map of the terrain.

The body of the book consists of three parts. Part 1 is the practical part: it focuses on the shift of the world we live in. The reality we experience is a substantially new reality. The challenge this "Macroshift" poses is that of constructive change, in us and around us, born of foresight empowered by awareness and understanding. We either change with our changing world—which we can do if we acquire the understanding and master the will—or we risk growing crises and ultimately breakdown.

Part 2 is the theoretical part, but it focuses on an eminently practical concern: how to understand the world we live in and the universe, its wider context. Not only our world, but science is also changing; the change there is in the form of a paradigm shift. The concept of reality emerging at the frontiers of scientific research has little if any resemblance to the classical concept we were taught in school. The new concept is broader—it extends to multiple universes arising in a possibly infinite meta-universe—and it is deeper, extending into dimensions below the domain of the quantum. It is also more inclusive, shedding light on phenomena that were ignored or considered "anomalous" and relegated to metaphysics, theology, or parapsychology but a few years ago.

The thirteen chapters that together make up part 1 (dealing with our changing world) and part 2 (outlining the changes in science's map of the world) make a coherent whole, but each chapter can also be read separately, as prompted by the concerns and interests of the reader. They are intended to help us understand our changing world as well as our changing map of the world and help us empower and guide our evolution as we move into the critical phase of today's Macroshift.

Part 3 moves from theory to hands-on practice. It describes the origins, the projects, and the principal objectives of the Club of Budapest, a global think tank founded by the author and dedicated to facilitating the changes that need to come about in our world by applying the insights of science's new map of reality to the cause of peace, sustainability, well-being, and human survival.

A closing section—the annex—breaks fresh ground in our scientific mapping of the deeper regions of human experience. It reviews a mind-boggling experience of the author and attempts to interpret it in light of the new map of reality. The experience ("transcommunication" with persons who have died recently) is of such staggering importance that it merits venturing beyond the bounds of established science—which, we should note, are by no means the bounds of human insight and understanding.

Part One

MACROSHIFT IN SOCIETY

ONE

Evolution or Extinction
THAT IS THE QUESTION

Had he lived today, Hamlet, Prince of Denmark, would affirm with deeper conviction than ever: To be or not to be is indeed the question. It is not the skull of an individual human being that Hamlet would ponder, but this living blue-green planet, the home of humanity. How long will it support us? Will we destroy its delicate balances, or will we set out to heal the damage we have already inflicted? Will we manage to evolve as a conscious social and cultural species—or will we become extinct like the dinosaurs?

The question is: Evolution or extinction?

A Chinese proverb warns, "If we do not change direction, we are likely to end up exactly where we are headed." Applied to today's world, this would be disastrous:

- There is deepening insecurity in countries both rich and poor and greater propensity in many parts of the world to resort to terrorism, war, and other forms of violence.
- Islamic fundamentalism is spreading throughout the Muslim world, neo-Nazi and other extremist movements are surfacing in Europe, and religious fanaticism is appearing the world over.
- Governments seek to contain violence through organized war-

fare; world military spending has risen for the past eight years running and has reached more than one trillion dollars a year.

- One in three urban dwellers in the world live in slums, shanty-towns, and urban ghettos. More than 900 million people are classified as slum-dwellers. In the poorest countries 78 percent of the urban population subsists under life-threatening circumstances.

- Although more women and girls are being educated than in previous years, in many parts of the world fewer women have jobs and more are forced to make ends meet in the "informal sector."

- Frustration and discontent continue to grow as both power and wealth are becoming further concentrated and the gap widens between the holders of wealth and power and the poor and marginalized populations. Eighty percent of the world's domestic product belongs to one billion people; the remaining 20 percent is shared by five and a half billion.

- Climate change threatens to make large areas of the planet unsuitable for human habitation and for an adequate level of food production. Very few countries are still food self-sufficient—and the internationally available food reserves are shrinking.

- The amount of available fresh water is diminishing rapidly; over half the world's population faces water shortages. On average, 6,000 children are dying each day of diarrhea caused by polluted water.

We are not heading in the right direction. Where do we go from here?

TWO SCENARIOS

1. No change: the BAU (Business As Usual) Scenario

Continuing in this way, the prelude to the inevitable breakdown manifests as critical conditions arising in the regions most directly exposed to the pernicious effects of climate change. In these regions, home to hundreds of millions of inhabitants:

- Changing weather patterns create drought, devastating storms, and widespread harvest failures.
- Coastal areas are flooded by rising sea levels.
- Famine spreads in areas dependent on adequate rainfall for food production and areas exposed to tornados, hurricanes, and violent storms.
- Massive waves of migrants from the worst-hit areas seek areas where resources are more assured.

The breakdown of the poorest and most directly exposed regions creates a global security threat:

- Epidemics of infectious diseases spread over Africa, Asia, and the Americas owing to heat waves, outbreaks of agricultural pests, and contaminated drinking water.
- The waves of migration to relatively well-off regions overload the local resource base and create conflict with the established populations.
- Terrorist groups, nuclear proliferators, narco-traffickers, and organized crime form alliances with unscrupulous entrepreneurs and expand the scale and scope of their activities.

On the way to breakdown we can anticipate drastic changes in economic and political processes and ecology, accompanied by military fallout.

The Economic and Political Processes

- Terrorism spreads, together with declared and undeclared attacks on countries suspected of harboring terrorists.
- The North Atlantic Alliance linking Europe, the United States, and Russia collapses.
- France, Germany, Russia, and China form a coalition to balance what they perceive as growing U.S. military-economic hege-

mony, joined by Brazil, India, South Korea, and other developing countries.

- Global military spending experiences a sharp rise, as the U.S. and its allies and the opposing bloc countries enter the spiral of an arms race.
- Global economic stagnation combined with U.S. unilateralism weakens the International Monetary Fund and the World Trade Organization. As regional economic agreements become more attractive than multilateral trade arrangements and bilateral trade with the U.S., trade wars become frequent and destabilizing.
- North–South trade agreements are cancelled and trade flows disrupted; the international economic/financial system is in shambles.
- People pressed into poverty join rebellions against local landowners and government officials.

The Ecological Dimensions

- Water and food shortages in Sub-Saharan Africa, China, southern Asia, and Mesoamerica generate water- and hunger-wars.
- The overexploitation of soils and overfishing of seas and rivers reduce yields in the industrialized countries and produce growing dependence on a shrinking stock of international food reserves.
- Starvation and unsanitary conditions accelerate the spread of HIV/AIDS, SARS, and other epidemics throughout the poor countries.
- The Gulf Stream vacillates, producing icy temperatures in spring and summer in western and northern Europe.

The Military Fallout

- Political and economic conflict between the U.S. and its allies, and the opposing military-political bloc reaches a crisis point; hawks and armaments lobbies press for the use of weapons of mass destruction.
- Strong-arm régimes come to power in the Southern Hemisphere, determined to use armed force to right perceived wrongs.

- Regional wars erupt in the traditional hot spots and spread to neighboring countries.
- The major military-political-economic power blocs decide to make use of their arsenals of hi-tech weaponry to achieve their economic and political objectives.
- Some among the new strong-arm régimes employ nuclear, chemical, or biological weapons to resolve regional conflicts.
- War fought with conventional and nonconventional weapons escalates to the global level.

No change leads to breakdown. But there is another path we could take.

2. *The Timely Transformation Scenario*
The First Steps

- The experience of terrorism and war together with rising poverty and the threats posed by a changing climate trigger positive changes in the way people think. The idea that individuals and small groups themselves can be effective agents of transformation toward a more peaceful and sustainable world captures the imagination of more and more people. People in different cultures and different walks of life pull together to confront the threats they face in common.
- The worldwide rise of popular movements for peace and international cooperation leads to the election of similarly motivated political figures, lending fresh impetus to projects of economic cooperation and intercultural solidarity.
- Political and opinion leaders wake up to the urgent need to come to the aid of the most immediately endangered populations and create a world-level organization to monitor the threats, provide warning, and raise the funds to undertake rescue operations.
- Local, national, and global business leaders decide to adopt a strategy where the pursuit of profit and growth is informed by the search for corporate social and ecological responsibility.

- An electronic E-Parliament comes online, linking parliamentarians worldwide and providing a forum for debates on the best ways to serve the common good.
- Nongovernmental organizations link up through the Internet and develop shared strategies to restore peace, revitalize war-torn regions and environments, and ensure an adequate supply of food and water. They promote socially and ecologically responsible policies in local and national governments and in business.

The Crystallizing Contours of a Cooperative World

- Money is reassigned from military and defense budgets to fund practical attempts at conflict resolution and the implementation of internationally agreed and globally coordinated social and ecological sustainability projects.
- A worldwide renewable energy program is created, paving the way toward a third industrial revolution that makes use of solar and other renewable energy sources to transform the global economy, provide clean water, and lift marginalized populations out of the vicious cycles of poverty.
- Agriculture is restored to a place of primary importance in the world economy, both for producing staple foods and for growing energy crops and raw materials for communities and industry.
- Business leaders the world over join forces in creating a voluntarily self-regulating eco-social market economy that ensures fair access to natural resources as well as industrial goods and economic activity to all countries and populations.

The Rise of a Sustainable Civilization

- National, continental, and global governance structures are reformed or newly created, moving states toward participatory democracy and releasing a surge of creative energy among empowered and increasingly active populations.
- The consensually created and globally coordinated eco-social market system begins to function; as a result the natural resources

required for health and well-being become available throughout the world community.

- International and intercultural mistrust, ethnic conflict, racial oppression, economic injustice, and gender inequality give way to a higher level of trust and the shared will to achieve peaceful relations among states and sustainability in the economy and the environment.

We could change direction: with a timely transformation we could create a peaceful and sustainable world. Will we create it? Einstein told us that we cannot solve a problem with the same kind of thinking that produced it. Yet, for the present we are still trying to do just that. We are fighting terrorism, poverty, criminality, cultural conflict, climate change, environmental degradation, ill health, even obesity and other "sicknesses of civilization" with the same means and methods that produced the problems in the first place—we are resorting to armies and police forces, technological fixes, and temporary remedial measures. We have not mustered the will and the vision to bring about timely transformation.

IS IT TOO LATE?

In the spring of 2006 the British biologist James Lovelock, who thirty years ago discovered that Earth possesses a planetary control system that keeps it fit for life (the "Gaia hypothesis"), proclaimed that this control system has been destroyed and will rapidly bring about conditions that may prove fatal for humanity. The heating up of the atmosphere through human activity will create, in Lovelock's words, "a hell of a climate." The average temperature will rise 14.4 degrees Fahrenheit in temperate regions and 9 degrees in the tropics. "The Earth's physical condition must be seen as seriously ill, and soon to pass into a morbid fever that may last as long as 100,000 years." "I think we have little option," Lovelock concluded in *The Revenge of Gaia*, "but to prepare for the worst, and assume that we have passed the threshold." The threshold he refers to is the point where the self-maintaining dynamic of the system breaks down and leads irreversibly to catastrophe.

Have we reached that catastrophic point already? We do not know for certain, but the news is not encouraging. The global climate is crashing, full of tipping points and feedback loops beyond which the slow creep of environmental decay gives way to sudden self-perpetuating collapse. Vital balances are degrading in the atmosphere, in the oceans and freshwater systems, and in productive soils. The consequences include the greenhouse effect and a reduction of the productivity of seas, lakes, rivers, and agricultural lands.

A number of critical processes feed on themselves and are out of control. As the Arctic ice melts, the sea absorbs more warmth, which makes for more melting; as Siberian permafrost disappears, the methane released from the peat bog below exacerbates the greenhouse effect and makes for more melting and thus for more methane.

But doomsday arguments miss a basic point: they do not recognize that not only is nature a dynamic system capable of rapid transformation but humanity is also. When such a system nears the point where the existing structures and feedbacks can no longer maintain its integrity, it becomes ultrasensitive and responds even to the smallest provocation for change. In this state "butterfly effects" are possible. (These effects are named after the butterfly-shaped "chaotic attractor" discovered by meteorologist Edward Lorenz as he attempted to map progressive change in the global weather. They are popularly identified with the idea that the tiny stream of air created by the flutter of the wings of a butterfly can amplify many times over and end by creating a storm on the other side of the planet.) In today's near-chaotic, unstable, and hence ultrasensitive world such "butterflies" as the thinking, the values, the ethic, and the consciousness of a critical mass in society can trigger fundamental transformation.

THE POSITIVE OUTLOOK

We are nearing a tipping point, but the situation is far from hopeless: near the threshold of systems-collapse, predictions of doomsday have a paradoxical effect. They raise people's level of awareness, motivate widespread consciousness change, and may end by becoming self-falsifying prophecies.

The political situation can turn paradoxical. Well-intentioned policies create the impression that the situation is in hand and the crisis is being managed, and thus they do not catalyze the will for fundamental transformation. A retrograde strategy is more useful in this regard. It inadvertently but effectively motivates people to insist on radical change; it catapults ever more people into action.

At the present time retrograde policies are still dominant. In the last analysis this is not a bad thing. In the more advanced segments of the population it raises the level of urgency of economic, social, and political reforms.

The Asian tsunami's carnage of innocent villagers and vacationers in South and Southeast Asia prompted worldwide acts of solidarity and generosity. The cataclysm produced by hurricane Katrina made people "find their feet" and march on Washington to protest the administration's policy of focusing on the oil war in Iraq to the neglect of preparedness for natural disasters and the plight of poor people at home. Will humanity wait for a natural or man-made catastrophe that kills hundreds of thousands or millions to come up with the will to change? It may then be too late. We must, and still can, head toward a timely shift in values, vision, and behaviors.

Evolution to a sustainable civilization, or descent into crisis, chaos, and possibly extinction: that, as Hamlet would now say, is the question.

TWO

Macroshift
THE DYNAMICS

We live in a crucial epoch—an epoch of instability and change. The future is open. We could go down in chaos and catastrophe, or pull ourselves up by our bootstraps to a peaceful and sustainable world. The choice between extinction and evolution is real. We need to understand how it came about and what it entails.

The first thing to understand is that the choice of destiny before us is not accidental: the way the world in which we live develops has a logic of its own. This logic is the logic of evolution, in nature as well as in society. Its hallmark is the alternation of periods of relative stability with epochs of increasing, and ultimately critical, instability. When instability reaches the critical point, the system either collapses or shifts to a new state of dynamic stability. These critical "tipping points" constitute Macroshifts, which involve all aspects and segments of society: the rich and the poor, the economic and the political systems, the private as well as the public sector.

We are approaching the threshold not only of a local or national but of a global Macroshift, driven by the cumulative impact of the unreflective use of potent technologies. Shortsighted power- and profit-hunger coupled with powerful technologies has triggered climate change, is producing famine and water scarcity, and is leading to coastal flooding as well as to a host of related and equally threatening processes in the ecology. Within the structures of civil society it is producing growing gaps between rich and poor, with attendant frustration, fundamentalism, and terrorism, triggering crime, violence, and war.

The threat of extinction is real, but it is avoidable. At the critical phase of a Macroshift fresh opportunities open, including the opportunity to evolve. In this case the opportunity is not to evolve genetically, for we are not merely a biological species, but to evolve socially and culturally, to a new society and a new culture—to a new civilization.

Evolution, whether in nature or in the human world, is characterized by certain basic features that recur independently of the nature of the things that evolve, and also of their particular time and place. The first of these recurrent features concerns the manner in which evolutionary processes unfold.

Wherever they occur, the processes of evolution are continuous and unrelenting, but not smooth and even. Aside from occasional temporary reversals, evolution is largely *irreversible,* and the way it unfolds is highly *nonlinear.* A seemingly enduring process of change suddenly forks off in a new direction. The systems become chaotic; more exactly, the kind of butterfly-shaped attractors that were discovered by meteorologist Edward Lorenz appear in the dynamic "portrait" of their evolution. As a result their trajectory forks off: it *bifurcates.* This process comes to the fore whenever and wherever complex systems undergo irreversible change.

At the threshold of a critical instability, fluctuations that were previously corrected by self-stabilizing negative feedbacks within the system run out of control—they break open the system's structure. The system enters a period of chaos. Its outcome is either the disintegration of the system into its individually stable components (*breakdown*) or rapid evolution toward a kind of system that is resistant to the fluctuations that destabilized the prior system (*breakthrough*).

At the critical bifurcation point various attempts are made to move beyond the crisis. Attempts to maintain the status quo are condemned

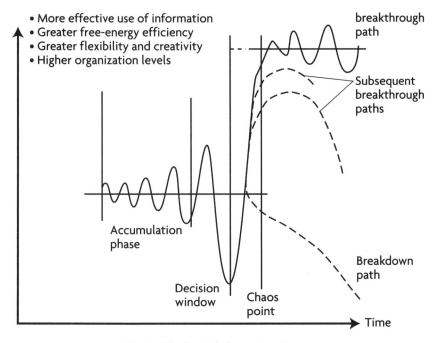

- More effective use of information
- Greater free-energy efficiency
- Greater flexibility and creativity
- Higher organization levels

breakthrough path

Subsequent breakthrough paths

Accumulation phase

Breakdown path

Decision window

Chaos point

Time

Fig. 1. The basic bifurcation diagram

to failure, and attempts to reach a new level of dynamic equilibrium do not always pan out. Some move the system toward a new equilibrium but prove unable to maintain it; they merely postpone the system's collapse. However, other attempts may be crowned by success. They conduce to a state in which the system has enhanced information-processing capacity and greater efficiency in the use of free energy as well as greater flexibility, higher structural complexity, and additional levels of organization (fig. 1).

EVOLUTION THROUGH BIFURCATION

Complex systems—biological as well as social—evolve through bifurcations. Evolution in the biosphere is an integral process; it encompasses unicellular organisms on the one end of the scale of organization and complexity, and entire biospheres populated by multicellular organisms on the other. The process is driven by the flow of free energy from the Sun. Free energy is transformed by plants into biomass; the biomass is consumed by herbivores that in turn are food for carnivores, creating a continuous cycle that constitutes an open thermodynamic system. This energy mill drives the biological and biochemical processes in the biosphere (fig. 2).

The evolutionary process is integral, but its unfolding is strongly nonlinear. Periodic bifurcations in the evolutionary history of biological and ecological systems mark the course of evolution on Earth, with its early phases occurring throughout the universe.

Evolution in the universe took off from physical systems and, through the chemical evolution of stars and related interstellar processes, moved progressively from the substratum of quarks and elementary particles to the atoms of the elements and the molecules and crystals formed by atoms. On Earth's energy-irradiated surface, evolution progressed fur-

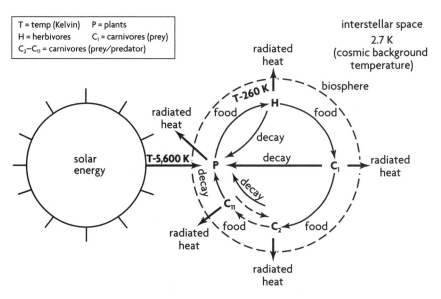

Fig. 2. The energy mill that powers life in the biosphere

The thermal energy gradient between the energy streaming from the Sun to the surface of Earth and the temperature of space around the planet (the cosmic background temperature) constitutes an energy mill where the heat energy of solar radiation is transformed into systems of increasing complexity and the waste energy—degraded to lower temperatures—is radiated off into space.

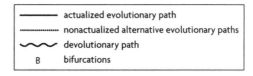

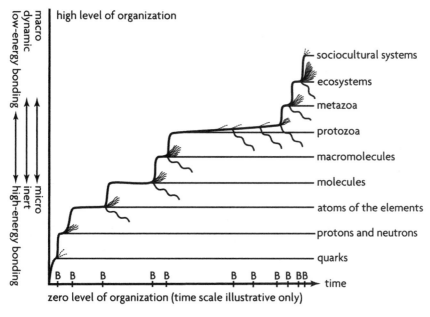

Fig. 3. *Evolution through bifurcations in nature*

ther. Solar radiation combined with submarine hot springs stirred the rich "molecular soup" in the shallow primeval seas of the young Earth and created progressively more complex structures: prokaryotic and then eukaryotic cells, then colonies of cells, and ultimately genuine multicellular organisms.

The evolutionary process on Earth produced entire species' lineages, moving in a continuous but nonlinear fashion to complex mammalian species embedded in multispecies local, regional, and continental ecologies. With the appearance of *Homo,* it encompassed the sociocultural and sociotechnological systems formed by human tribes and communities (fig. 3).

Evolution through periodic bifurcations gave birth to the lineage of hominids. The family of primates split off from the then existing species of mammals around 40 million years ago. The first primates were the

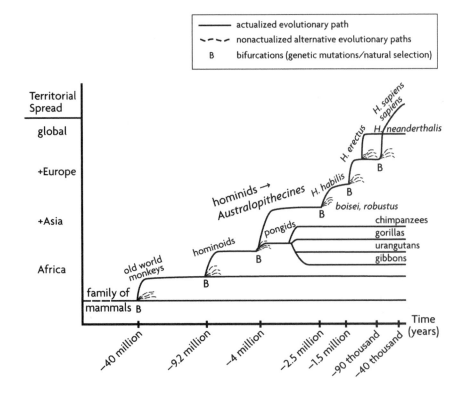

Fig. 4. The evolutionary path to Homo

old world monkeys that populated wide areas of Asia and Africa. Then, about 9.2 million years ago the primate family split into two groups. One, the pongids, stayed with arboreal life and, while several branches became subsequently extinct (such as *gigantopithecus* and *sivapithecus*), the survivors evolved into the modern apes: the chimpanzees, gorillas, orangutans, and gibbons. The other group became terrestrially based bipedalists: the family of hominids.

About 4 million years ago the early hominid *Australopithecines* were widely distributed in Eastern and Southern Africa. Living in small bands, they managed to survive the dangers of terrestrial life. Around 2.5 million years ago they split into different branches. A branch that became extinct included numerous subspecies, such as *boisei* and *robustus*, while the surviving branch led to *habilis* and *erectus*, and ultimately to *sapiens* (fig. 4).

Although the details of hominid evolution are not definitively established, it appears that modern human beings, *H. sapiens sapiens,* first evolved in Africa. Some forty thousand years ago *sapiens* appeared in Europe, probably co-inhabiting the continent with *H. neanderthalis.* The latter disappeared around thirty thousand years ago, making *sapiens sapiens* the sole survivor of the hominid branch.

SOCIAL EVOLUTION THROUGH MACROSHIFTS

With *sapiens sapiens* evolution shifted from the biological to the socio-cultural-technological domain. In this domain it is not the genetic structure that mutates but the dominant civilization: how people are organized, what ideas and values they entertain, and how they see themselves and the world around them. Mutations in society are all-encompassing, involving every segment and every aspect. They are shifts in civilization: shifts that are "macro." Across numerous hills and valleys, and occasional abrupt leaps, these Macroshifts drive toward the progressive integration of different peoples, enterprises, economies, societies, and cultures in systems of larger and larger dimensions.

The evolution of human groups in intercommunicating kinship or social structure-based communities is described in the chronicles of history. This is a complex process, for human beings are not simply the passive subjects of evolution but are active (even if usually not voluntary and conscious) agents that influence its unfolding. Nevertheless, even if they do not will it, or even know it, the societies formed by human beings undergo an evolutionary process that is analogous to that which occurs in biological nature. In history, too, bifurcations intersperse comparatively stable periods and lead to systems that are more and more complex and are further and further from inert states of entropy and thermodynamic equilibrium.

The evolution of human societies has been driven by the innovations that periodically destabilize the existing systems. Major innovations have been rendered possible by *sapiens'* capacious cranium, harboring a brain of some 1,350 cm^3. This enabled our forebears to develop an expressive and then a symbolic language, conceptual thinking, advanced tool use, and group behavior based on the cooperative use of progressively more sophisticated technologies.

At first, societal evolution was slow: Paleolithic Stone Age societies were highly enduring, with a low level of innovation and great stability. The first major innovation that rocked these societies was the domestication of plants and animals around ten thousand years before our time: the "Neolithic Revolution." It transformed nomadic hunter-gatherers into settled pastoralists, and then into agriculturists. At that time and afterward, throughout history, bifurcations were triggered by advances in the technologies devised by human groups. Technological innovations included the control of fire, the invention of the wheel, the design of progressively more sophisticated tools, and the invention of more and more powerful devices for extending the power of human muscle and the human brain. Such innovations enabled humans to live in larger and larger communities, with progressively greater social differentiation and divisions of labor.

Following the discovery of how to ignite, conserve, and transport fire, the paramount innovation was pastoralism and the early forms of agriculture. Subsequent innovations—including the invention of the alphabet and the number system, the means of communication over vaster distances, and the stratification of societies from the tribal circle of elders to the hierarchically organized state—transformed groups of Neolithic pastoral-agrarian communities into the vast archaic empires of Babylonia, Egypt, India, and China.

Less than four thousand years ago at the rim of the Mediterranean there was another major bifurcation: in classical Greece nature philosophers pioneered a societal mutation that replaced mythical concepts with theories based on observation and elaborated by reasoning. Greco-Roman civilization entered the scene of history. The pre-Socratic philosophers evolved the "heroic mind," present in Homer and the early epics, into the visionary and the theoretical mind and then the rational mind epitomized by Plato and Aristotle. *Logos* became the central concept: it was at the heart of philosophy as well as of religion. Together with the

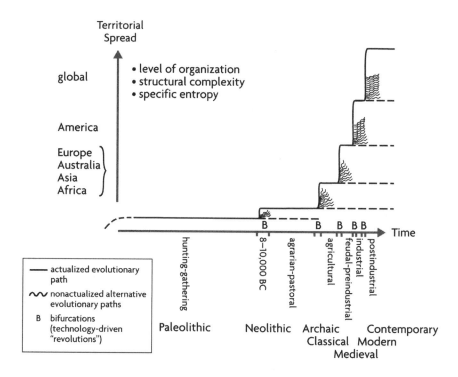

Fig. 5. The path of human sociocultural and sociotechnological evolution

valuation of quantitative measurements, it provided Western civilization with the rational foundation upon which it was to build for nearly two and a half thousand years (see fig. 5).

After the fall of the Western Empire of Rome and the founding of the Byzantine Empire in 476 CE, a further shift occurred in the development of European societies. The rise of Christianity modified the classical culture of Greece. The medieval belief system added to the classical concepts a divine source: the world's creator and prime mover as well as ultimate judge. Reason came to be embodied in the Holy Trinity and incarnated in man, God's creation. This belief system, whose principal elements were elaborated by St. Augustine and Thomas Aquinas, was dominant in European civilization until the advent of the modern age.

The rationality of the Greeks, borrowed and elaborated by the

Romans, was conserved in medieval fiefdoms and princedoms, notwithstanding the addition of Christian elements. It found expression in the creation and use of mechanical devices such as clocks, windmills, watermills, animal-drawn agricultural implements, and horse-drawn carriages.

A further shift occurred in the sixteenth and seventeenth centuries. Although medieval Europe's culture was otherworldly and Christian, in everyday practice it was mechanically colored; it embraced the concept elaborated by Giordano Bruno and Galileo Galilei: the world as a giant machine. This concept, underpinned by new scientific discoveries and wedded with traditional handicrafts, led to an entire series of technological innovations. These included the harnessing of the power of steam and later oil and the invention of mass production for mass markets. Europe, followed shortly by America, entered the industrial age.

Thanks to an accelerating series of ever more powerful technological innovations *sapiens* became the dominant species on the planet. But this reign is not assured. In its present form, industrial civilization is not sustainable. In the opening years of the twenty-first century the industrial age is shifting into a post-industrial age, impelled by the "second industrial revolution"—a revolution hallmarked by the advent of the technologies of information and communication. These technologies are more powerful than the steam- and fossil fuel–based technologies of the first industrial revolution, and the "revolutions" they catalyze are unfolding much faster than the first industrial revolution: in a matter of years instead of decades or centuries.

In the past Macroshifts were local, national, or regional. Today's Macroshift is global. Humanity's societal evolution has reached the dimensions of the planet.

THE PHASES OF A MACROSHIFT

Macroshifts have recognizable phases. Typically there are four major phases: first an initial phase of gradual but ongoing change, then a subsequent phase of more rapid build-up. After that comes a phase of crisis and bifurcation, and ultimately a concluding phase that can be one of breakthrough to a new and more stable system or breakdown into crisis and chaos (see fig. 6).

1. The Trigger Phase

Innovations in "hard" technologies (tools, machines, operational systems) bring about greater efficiency in the manipulation of nature for human ends.

2. The Transformation Phase

Hard technology innovations irreversibly change social and environmental relations and bring about, successively,

- a higher level of resource production,
- faster growth of population,
- greater societal complexity, and
- growing impact on the social and the natural environment.

3. The Critical (or "Chaos") Phase

Changed social and environmental relations put pressure on the established culture, placing into question time-honored values and worldviews and the ethics and ambitions associated with them. Society becomes chaotic in the chaos theory sense of the term: it does not lack order but exhibits a subtle order that is extremely sensitive to fluctuations. The evolution of the dominant culture—the way people's values, views, and ethics respond to change—determines the outcome of the system's chaos leap, that is, the way its developmental trajectory forks off.

4(a). The Breakdown Phase

The values, worldviews, and ethics of a critical mass of people in society are resistant to change, or change too slowly, and the established institutions are too rigid to allow for timely transformation. Social complexity, coupled with a degenerating environment, creates unmanageable stresses. The social order is exposed to a series of crises that soon degenerate into conflict and violence.

4(b). The Breakthrough Phase

The mind-sets of a critical mass of people evolve in time, shifting the culture of society toward a better-adapted mode. As these changes take hold, the improved social order—governed by more adapted values, worldviews, and ethics—establishes itself. Society stabilizes itself in its changed condition.

Just as in nature, the first phase of society's Macroshift is triggered by the destabilization of the previously dominant system. But, unlike in nature, in the human world this occurs not because of external factors (such as the emergence of other species or other changes in the milieu) but through internal, self-created changes—changes in the way people

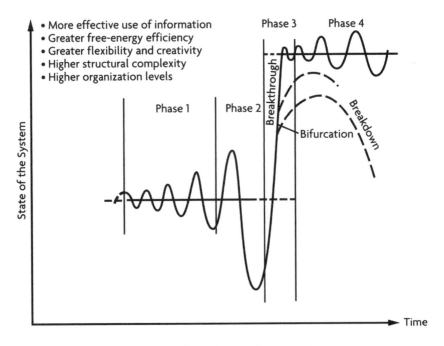

Fig. 6. The four phases of a Macroshift

The four phases of a Macroshift can be mapped into the basic bifurcation diagram. In the first phase society's stability is rocked by a mounting series of fluctuations. In the second phase the fluctuations exceed the ability of society's structures and institutions to govern and manage. In the third phase the Macroshift builds toward a point of bifurcation where society's evolutionary path forks off. The Macroshift then enters the fourth and final phase.

relate to each other and to their environment. For the most part these changes are driven by innovations in the dominant technologies: in the technologies of human interaction and of the exploitation and manipulation of nature.

For an established social order, technological innovations are a mixed blessing. On the one hand innovations in the "hard" (matter- and energy-transforming) technologies help people do what they want to do with greater ease and less investment of time, energy, and money. They amplify the power of muscles to move and transform matter, they extend the power of the eye to see and the ear to hear, and they enlarge the power of the brain to register and compute information. On the other hand new technologies have unforeseen and often seriously undesirable "side effects." They place in question established practices, institutions, and ideals. They change people's ways of working and living, and they have a negative impact on the environment.

A time comes when the accumulation of innovative hard technologies exceeds the ability of society's structures and institutions to govern and to manage. Resource production increases, both through a more effective exploitation of the already exploited resources and by opening up new resources—for example, coal in addition to wood, then oil in addition to coal. A larger quantity and a wider variety of resources enable more people to produce and to consume. As a result, the population grows. But a larger population using more, and more kinds of, resources cannot make do with the kind of structures that served life based on simpler technologies and more limited resources. There is a need for special skills and special-purpose organizational structures. As these are developed the complexity of society also grows, compounding the instability created by the growth of its population and its resource base.

As more people exploit more resources through more sophisticated technologies dependence is created between previously separate people and economies. The range of interaction expands: there are increasing exchanges between diverse and far-flung societies and cultures. This puts pressure on society's institutions and governance structures; new ways of governing and administering communities and doing business are required. Some people and some societies come up with new ways and reap benefits; others fail to come along. Society polarizes into modern and traditional, rich and poor, and powerful and marginalized segments.

A complex society using sophisticated technologies creates another unexpected side effect: it places a critical load not only on society's institutions and governance structures but on the environment. Nature suffers in unforeseen ways: forests fail to regenerate, soils are impoverished, the climate changes, freshwater tables are lowered, ocean levels rise, and the atmosphere becomes polluted.

Ultimately the Macroshift builds toward a point of bifurcation, the critical third phase at which society's evolutionary path is rapidly decided. As in nature, bifurcations in society are triggered by instabilities that are beyond the ability of the system to overcome: this is the true meaning of "unsustainability." The status quo becomes untenable, and the system either comes up with new ways of maintaining itself or it goes under.

In contemporary society unsustainability is the result of economic and social expansion combined with environmental change and degradation. These produce inequalities and imbalances that disorient people and overload the administrative and control capability of society's dominant structures and institutions. Frustration breeds resentment and generates hate and violence. Society enters a period of social and political crisis.

Then comes the Macroshift's final phase. The bifurcation of society's evolutionary trajectory is launched by energy- and material-transforming "hard" technologies but it is not decided by them. In deciding the outcome of a bifurcation, the critical factors are information-processing "soft" technologies that are knowledge-intensive and value-sensitive. These are the technologies of social, economic, and political organization, and they express and define the civilization that dominates society. Depending on the evolution of the dominant civilization, society either breaks down in violence and chaos or breaks through to a more adapted sustainable civilization.

This four-phase process is the reality of the world we live in. Two of the phases are behind us, and the last phase is ahead of us. We live in the third phase: the critical phase of a societal bifurcation. The challenge is not to enter phase 4(a), the breakdown phase, but phase 4(b), the break-through phase (fig. 7).

NAVIGATING THE GLOBAL MACROSHIFT

In the human world, unlike in nature, a bifurcation can be decisively influenced by conscious will and considered purpose. Human will and purpose decide whether the world heads toward breakdown or toward breakthrough. This sensitivity to human intervention is a remarkable feature of today's civilization. It places a unique opportunity in our hands: the opportunity to tip the scales of human destiny.

We need a deeper understanding of the direction of evolution through Macroshifts. Concepts coming from the systems sciences can provide the necessary insights. We begin with the concept of "suprasystem." The formation of higher-level systems through the interlinking of previously more autonomous systems (which become the emerging system's sub-systems) is a familiar notion in general systems and general evolution theory. Suprasystems emerge through the creation of "hypercycles" in

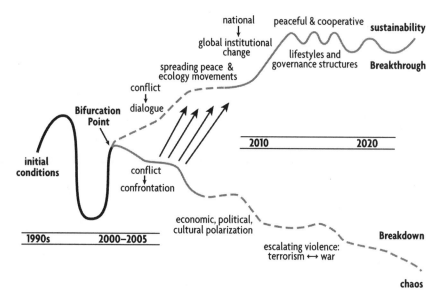

Fig. 7. The current bifurcation point and the requirement for a timely shift from the breakdown to the breakthrough path

which the subsystems are linked by cycles that mutually catalyze each other: so-called cross-catalytic cycles. The result is that the subsystems become increasingly interdependent, and the suprasystem jointly constituted by them takes on structure and autonomy.

This process is evident in today's world, most visibly in the world of business. A given enterprise—or a business unit of an enterprise—maintains itself by auto-catalytic cycles: it replenishes its human and natural resources, renews its infrastructure, and generally reproduces enough of itself to remain operational. Such enterprises, or business units, also develop links with other enterprises and business units in their industry environment. This produces a wider system ("suprasystem") that is viable and sustainable, provided that its principal units remain viable and sustainable. The requirement is that each business unit (or enterprise) both renew and maintain itself, and produce at the same time a viable business environment for the other units (or enterprises). In this

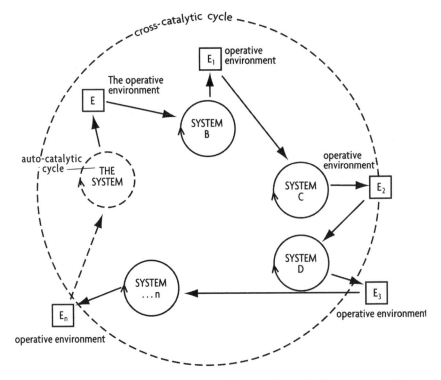

Fig. 8. Cross-catalytic and auto-catalytic cycles in the business world

manner enterprise or business unit A maintains itself and produces an operative environment for enterprise or business unit B. Enterprise or business unit B maintains itself and produces an operative environment for enterprise or business unit C—and so on. Ultimately the cycle closes on itself, as enterprise or business unit N (or the whole set of enterprises and business units that comprises the system) produces a viable operative environment for enterprise or business unit A. This constitutes a cross-catalytic cycle, the basis for the functioning of the suprasystem that is the enterprise itself or the multi-enterprise group (see fig. 8).

Suprasystem formation through cross-catalytic cycles also takes place on the political level. The over 190 nation-states of today's world claim independence and sovereignty for themselves, but their autonomy is increasingly restricted: for the most part nation-states are highly dependent on other states and on transborder economic and financial flows. National governments no longer have sovereignty over their own territory: all manner of weapons penetrate their frontiers, the same as information, goods, and people. Given that international and transnational exchanges of goods and technologies are in the hands of transnational corporations, national governments are unable to ensure the viability of the nation's economy, and given that pollution knows no frontiers, they are likewise unable to guarantee the physical integrity of their territory.

Even if they are large and wealthy, nation-states cannot survive in isolation; a condition of their self-maintenance is that they produce viable conditions for the states with which they are economically and politically linked. This constitutes a supranational cross-catalytic cycle. It is the basis for the functioning of the transnational organizations of which the European Union (EU) is a prime example.

In today's Macroshift the suprasystem-building process leads from national to regional level systems, and ultimately to a system on the global level. The process is evident, and it is accelerating. Regional economic groupings such as the EU and ASEAN (Association of Southeast Asian Nations) operate in conjunction with global intergovernmental organizations, such as the World Bank group and the World Trade Organization, and take over more and more of the functions of national governments. Control of the flow of goods, information, and people shifts toward the global level. At the same time nationally based industries evolve a transnational and even global dimension, and both global corporations and intergovernmental organizations become key actors on the contemporary scene—primary subsystems in the emerging planetary economic and political suprasystem (see fig. 9).

A NOTE ON THE GLOBAL CORPORATION

The global business corporation is a potent and remarkable but not an arbitrary phenomenon: it is the outcome of a coherent evolutionary process. This particular process has been spurred by two developments. One is the intensification of transborder flows of matter, energy, and information, and the other is the decoupling of the private sphere of business from the public sphere of the nation-state. Transborder flows of matter, energy, and information became significant factors in the evolution of business in the course of the twentieth century, whereas the separation of the private from the public sphere has been a characteristic of the modern nation-state almost from its birth in the period following the Peace of Westphalia. The institutions of the nation-state were decoupled on the one hand from the spiritual sphere, manifested in the separation of State and Church, and on the other from the structures that arose as citizens pursued their own economic interests. The latter process resulted in the separation of the public and the private sectors of society.

Until the second half of the eighteenth century, the economic activities of the citizens of nation-states were largely limited to local resources and locally developed skills. Resources were mostly local, for traditional technologies used primarily what was available in the immediate environment. Until the late eighteenth century few technological innovations called for international searches for additional energies and materials. Basic agricultural tools, such as the sickle, the hoe, the chisel, the saw, the hammer, and the knife, continued in use substantially unchanged. Significant innovations occurred only in agriculture, with the introduction of new irrigation technologies and new varieties of plants.

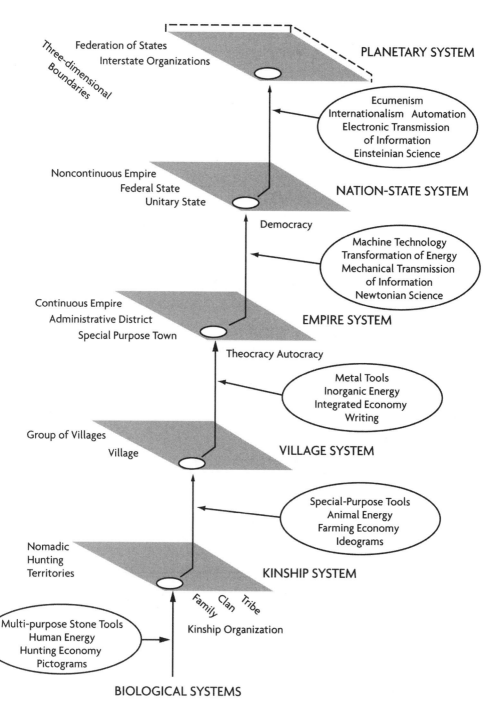

Fig. 9. Major stages and technology-driven transformations in society's evolutionary path toward the global level

The internationalization of economic activity began when traditional handicrafts came in contact with classical physics and its offshoots, classical chemistry and classical thermodynamics. In 1698 Thomas Savery patented the first crude steam engine based on Denis Papin's 1679 Digester, an early form of pressure cooker. With Thomas Watt's commercially viable steam engine, the power of steam was effectively harnessed. The applications of chemistry followed shortly thereafter. In the late eighteenth century the development of synthetic dyes led to the emergence of entire chemical industries. The consequences were felt first in textiles when innovations in spinning cotton stimulated inventions leading to machines capable of factory-based mass production. Industrial development then spread from textiles to iron, as cheaper cast iron replaced more expensive wrought iron. The size of the businesses that utilized these technologies increased, and so did their appetite for energy and material resources.

By the second half of the nineteenth century the first industrial revolution brought a battery of new technologies on the scene. The Bessemer steel process was developed in the 1860s, and the rotary kiln used in cement production was patented by Fredrick Rancome in 1885. Many of the twentieth-century technologies in the automobile, steel, cement, petrochemical, and pharmaceutical industries were spawned in the same period. The traction-based combustion engine appeared in the 1880s, simultaneously with Edison's electric light bulb, followed shortly by Marconi's wireless telegraph machine and the Wright brothers' flying machine.

In the course of the twentieth century technological innovations shifted industrial production from coal and steam, textiles, machine tools, glass, pre-Bessemer forged steel, and labor-intensive agriculture, to electricity, the internal combustion engine, organic chemistry, and large-scale manufacturing that was soon to grow beyond the borders of national states. The shift was intensified when in the latter part of the century the second industrial revolution replaced reliance on massive energy and raw material inputs with the intangible resource known as information. In the last quarter of the twentieth century vast quantities of information came to be stored on optical disks and communicated by fiber optics and networked computers. New information technologies rationalized and dropped the costs of production and consumption and led to vast increments in the mining, production, use, and eventually discarding of a wide variety of manufactured goods.

The new technologies intensified the transborder flows of information, energy, raw materials, and finished products. They brought into play the natural and human resources of whole continents, ultimately the entire globe. This development required successful businesses to move beyond the borders of the countries in which they had originally operated. The operational units and technological installations of innovative corporations spread to all parts of the globe.

Owing to the simultaneous presence of global resource flows and the separation of the private and the public spheres, businesses that started life with local labor and local resources were able to grow into globally extended corporations. Today some global corporations are wealthier than all but a handful of nation-states, and the largest among them employ more people than the population of the majority of the world's nearly 200 nation-states.

IN CONCLUSION

Societal evolution is a long-term process, with roots extending back to our species' prehistory. It is on the whole irreversible, and it is nonlinear, beset with periodic bifurcations. The current bifurcation takes human community-building from the nation-state to the planetary level. It is as profound as any evolutionary process in history, but it is incomparably faster than anything that went before. It poses enormous challenges of adaptation for individuals and societies.

The principal challenge facing people and societies today is to shift the civilization that now dominates human life and determines its future. This, as we shall discuss in chapter 7, means a shift from *Logos*, today's economically, politically, and culturally fragmented civilization, to *Holos*, a global civilization that possesses the will and the vision to achieve solidarity and translate it into international and intercultural coexistence and cooperation (fig. 10).

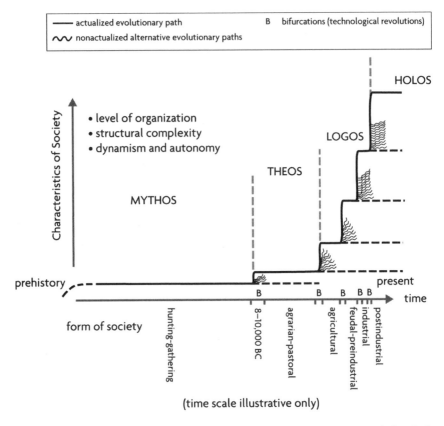

Fig. 10. Key features of the dominant culture of past civilizations and the challenge of shifting to a culture capable of ushering in a global civilization

THREE

The Roots of Unsustainability

In the first decade of the twenty-first century, humanity is still dominated by the materialistic, energy- and resource-intensive, and narrowly self-concerned technological civilization born in the West and extended to all continents. A linear continuation of the trends and processes engendered by this civilization is not sustainable; it would provoke major crises and ultimately breakdown.

Before getting down to the question of how a better civilization could emerge, we should look at the roots of unsustainability in the presently dominant civilization—roots that reside in the condition of civil society, in the workings of the economic and the financial system, and in the state of the global ecology.

CIVIL SOCIETY

The lion's share of current world population growth occurs in the developing countries. As a result—unless starvation and inhuman living conditions decimate those populations—the centers of poverty will expand dramatically. The population of the least developed countries will increase from 800 million today to 1.7 billion in 2050, with populations tripling in Afghanistan, Burkina Faso, Burundi, Chad, Congo, East

Timor, Guinea-Bissau, Liberia, Mali, Niger, and Uganda. On the other hand, the population of the industrialized countries will either shrink or remain constant.

In the poor countries the exigencies of economic survival are destroying the traditional extended family. As women are obliged to leave the home in search of work, poverty breaks apart even the nuclear family. Women are extensively exploited, given menial jobs for low pay. Children fare even worse. According to the International Labour Organization, fifty million children worldwide (mostly in Africa, Asia, and Latin America) are presently working. They are employed for a pittance in factories, in mines, and on the land, and many more are forced to venture into the hazards of life on the street as "self-employed vendors"—or plain beggars.

The unsustainability of conditions in civil society is not the consequence solely of economic gaps and imbalanced patterns of population growth; under growing stress, social structures are breaking down in rich countries as well. Increasingly, men and women find more satisfaction and companionship at work than at home. After children have "flown the nest," it is becoming usual for couples to seek fulfillment with other partners rather than restructuring the family relationship in a childless home. In the U.S. the rate for first marriages ending in divorce is 50 percent, and about 40 percent of children grow up in single-parent families for at least part of their childhood. Families eat meals together less and less frequently, and when they do, the TV is likely to be the center of attention. Children's media exposure to TV, video games, and "adult" themes is increasing, and exposure to such imagery, researchers find, connects with violent and sexually exploitive behavior. Teens face the peer challenge of "freer" sex, where loose "hooking up" for one-night stands is coming to be seen as normal, and building deep emotional relationships with sexual partners is considered out of date.

Many of the functions of family life are taken over by outside interest groups. Child rearing is increasingly entrusted to kindergartens and company or community day-care centers. The provision of daily nourishment is shifting from the family kitchen to supermarkets, prepared-food industries, and fast-food chains, and leisure-time activities are strongly colored by the marketing and public relations campaigns of commercial enterprises.

THE ECONOMIC SYSTEM

Economic growth continues on the global level, but its benefits accrue to ever fewer people. Hundreds of millions live at a higher material standard of living, but thousands of millions live in shantytowns and urban ghettos. Without significant change, by the middle of this century some 90 percent of the world's people will live in the poor countries, and the great majority of states themselves will be poor.

The unsustainability of the current distribution of wealth threatens the life and survival of the poor and poorest populations. The threat is not due to the finiteness of the planet's physical and biological resources but to the imperfect functioning of the economic system that processes and distributes them. With better distribution of the resources, the entire human population could enjoy a decent standard of living. For example, if food supplies were more equitably distributed, every person could receive about a hundred calories more than the 1,800 to 3,000 calories required for health. But currently people in North America, Western Europe, and Japan use (and to a large extent waste) 140 percent of their daily caloric requirement, whereas populations in countries such as Madagascar, Guyana, and Laos obtain merely 70 percent.

Current trends in energy consumption are likewise unsustainable. The average amount of commercial electrical energy consumed by Africans is half a kilowatt-hour (kWh) per person, the corresponding average for Asians and Latin Americans is 2 to 3 kWh, and for Americans, Europeans, Australians, and Japanese it is 8 kWh. The average American burns five tons of fossil fuel per year, in contrast with the 2.9 tons by the average German. The average American places twice the environmental load on the planet as the average Swede, three times that of the Italian, thirteen times the Brazilian, thirty-five times the Indian, and two hundred and eighty times the Haitian.

THE FINANCIAL SYSTEM

The world's financial system is unsustainable as well. This unsustainability is structural and not immediately apparent; despite periodic crises, economic growth, measured in monetary terms, continues.

Yet the current patterns of economic growth unbalance the financial system. The U.S. has a growing trade deficit: the value of the goods it imports is far above the value of the goods it exports. The opposite is the case in China and other Asian economies: they have a growing trade surplus, as the value of their exports is consistently above the value of their imports. At present, the Asian economies are financing U.S. overspending, but not voluntarily: central banks with large foreign exchange reserves, like China, Japan, and other Asian countries, are captives of America's fiscal policy. This trend cannot unfold indefinitely: the financial imbalance is growing toward untenable dimensions.

The International Monetary Fund's *Economic Outlook* noted in 2005 that it is no longer a question *whether* the world's economies will adjust, only *how* they will adjust. If measures are unduly delayed, the adjustment could be "abrupt," with hazardous consequences for global trade, economic development, and international security. For the time being the adjustment is postponed, for it involves painful losses on the part of the reserve-currency economies and a major and likewise painful adjustment in U.S. economic and trade policies.

THE ECOLOGICAL SYSTEM

The way humanity exploits the environment is intrinsically unsustainable. In regard to both the physical and the biological resources of the planet, the rising curve of demand is exceeding the descending curve of supply. This is historically unprecedented. For most of history, humanity's demand has been insignificant in relation to the available resources. But in the six decades since World War II, humanity has consumed more of the planet's physical and biological resources than in all of history prior to that time. Global consumption is nearing planetary limits. The production of oil, fish, lumber, and other major resources has already peaked, half the world's forests and 40 percent of the coral reefs are gone, and annually about 23 million acres of forest are lost.

It is not the sheer size of humanity that is the cause of the problem but its per capita resource use—it is out of proportion to its size: 6.4 billion humans are only 0.014 percent of the biomass of life on Earth and 0.44 percent of the biomass of animals. The disproportion is shown by quantitative measures of human resource use, such as the "ecological footprint" (the measure of the area of land required to support a person or a population). The proportionate level of agricultural production—where production is sustainable—is a footprint of 4.2 acres per person. However, the global average is a footprint of 7 acres. The extremes range from 1.23 acres in Bangladesh to 25.5 acres in the United States and in the oil-rich Arab countries.

The crossing of the curves is accelerated by the progressive impairment of ecological balances, which reduces the available supplies. According to former UN Assistant Secretary-General Robert Muller, each minute 52 acres of tropical forest are lost, 50 tons of fertile topsoil are blown off, and 12,000 tons of carbon dioxide are added to the atmosphere. Each hour 1,693 acres of productive dry land become desert, and each day 250,000 tons of sulfuric acid fall as acid rain in the Northern Hemisphere. An estimated 100,000 chemical compounds are injected into the land, rivers, and seas, millions of tons of sludge and solid waste are dumped into the oceans, and billions of tons of CO_2 are released into the air.

The situation as regards water is critical. In the past, the planet's available freshwater reserves were adequate to satisfy human needs: in 1950 there was a potential reserve of nearly 17,000 m³ of freshwater for every person then living. Since then the rate of water withdrawal has been more than double the rate of population growth, and consequently in 1999 the per capita world water reserves decreased to 7,300 m³. If current trends were to continue, in the year 2025 there would be only 4,800 m³ of reserves per person. This amount, combined with the uneven access to the reserves, would create major health hazards in many parts of the world. Already today about one-third of the population does not have access to sufficient supplies of safe water, and by 2025 two-thirds of the population will live under conditions of extreme water scarcity. Whereas Europe and the United States will have half the per capita water reserves they had in 1950, Asia and Latin America will have only a quarter. The worst hit countries will be in Africa, the Middle East, and south and central Asia. Here the available supplies may drop to less than 1,700 m^3 per person.

The trend concerning the availability of productive land is likewise critical. The Food and Agriculture Organization (FAO) estimates that on the global level there are 7490 million acres of high quality cropland available, 71 percent of it in the developing world. This quantity is decreasing due to soil erosion, destructuring, compaction, impoverishment, excessive desiccation, accumulation of toxic salts, leaching of nutritious elements, and inorganic and organic pollution owing to urban and industrial wastes. In some parts of the world, this augurs major food shortages. China has a population that is five times that of the United States but has only one-tenth as much cultivated land; it is feeding 24 percent of the world's population on 7 percent of the world's agricultural land. This small percentage is further diminishing. Due to urban sprawl and the construction of roads and factories, 37 million acres of China's cultivated land have already been converted to nonagricultural use. Of the remaining 247 million acres one-tenth is highly polluted, one-third is suffering from water loss and soil erosion, one-fifteenth is salinized, and nearly 4 percent is in the process of turning into a desert.

Worldwide, 12 to 17 million acres of cropland are lost per year. If this process continues, some 741 million acres will be lost by mid-century, leaving 6.67 billion acres to support 8 to 9 billion people—no more than 0.74 acre per person, the area for subsistence-level food production.

The pollution of the atmosphere is another unsustainable trend. The amount of air that humans, and even all organisms taken together, need is minuscule compared to the size of the atmosphere that surrounds the planet. But here, too, it is a question of quality rather than quantity. Polluted air and air of inadequate oxygen content are of little use. Yet the oxygen content of the atmosphere is progressively being reduced, and its carbon dioxide and other greenhouse gas content is being rapidly increased. Since the middle of the nineteenth century oxygen has decreased mainly due to the burning of coal; it now dips to 19 percent of total volume over impacted areas and to 12 to 17 percent over major cities. At 6 or 7 percent of total volume, life can no longer be sustained. At the same time the share of greenhouse gases is growing. Two hundred years of burning fossil fuels and cutting down large tracts of forest has increased the atmosphere's carbon dioxide content from about 280 parts per million to over 350 parts per million.

The influx of gases from human activity is paralleled by the growing influx of gases from nature, which is now largely triggered by human activity. In Siberia, an area of permafrost spanning a million square kilometers, the size of France and Germany combined, has started to melt for the first time since it formed at the end of the last ice age 11,000 years ago. Russian researchers found that what was until recently a barren expanse of frozen peat is turning into a broken landscape of mud and lakes, some more than a kilometer across. The area, the world's largest frozen peat bog, has been producing methane since it formed at the end of the last ice age, but most of the gas has been trapped under the permafrost.

The west Siberian peat bog may hold as much as 70 billion tons of methane, a quarter of all of the methane stored in the ground around the world. Calculations show that the melting peat bog could release around 700 million tons of carbon into the atmosphere each year, about the same amount that is released annually from all of the world's wetlands and agriculture. This would double atmospheric levels of the gas, leading to a 10 percent to 25 percent increase in global warming. Changes in the chemical composition of the atmosphere trigger alterations in the climate. Climate change has already reached the danger point.

Temperatures in the western Arctic are at a 400-year high, and in September of 2005 satellite pictures testified that the extent of the Arctic ice cap is 20 percent below the long-term average for the month of September. If this trend continues the Arctic Ocean will be completely ice free before the end of the century and perhaps before. This is a realistic prospect, since the warming process feeds on itself: as ice disappears, the surface of the sea becomes darker, absorbing more heat. Less ice forms, which means that the sea becomes still darker, absorbing still more heat.

The progressive reduction of the Arctic ice cap is altering the world's weather. It first of all threatens Europe, as the volume of fresh water streaming into the North Atlantic may ultimately deflect the Gulf Stream. That would flood Western Europe with frigid waters, creating winters of Siberian cold over England and much of the continent.

While Europe is threatened with a colder climate, most of the planet will be subjected to rising temperatures. If nothing decisive is done to deflect the warming trend, the damage to the Amazon rain forest, already apparent, will become irreversible. There will be widespread destruction of coral reefs, the alpine flora of Europe and Australia will disappear, and there will be severe losses of China's broad-leaved forests.

Climate change will play havoc with the yield of agricultural lands and thus threaten entire human populations. Although in cold regions with short growing seasons yields could increase, they will decrease in tropical and subtropical areas where crops are already growing near the limit of their heat tolerance. The negative effects outweigh the few positive consequences. Climate experts estimate that as the Gulf Stream deflects, the North Atlantic/Northern Europe region will become generally cooler, with storms and rain concentrated over Siberia. Most of the Southern Hemisphere will become warmer and dryer; even the monsoon may discharge its torrential rains over the sea rather than over land. Evidently, such changes are a threat to the food supply of the entire world.

The recognition that climate change–induced dangers are real and need to be combated is growing. A new milestone was reached in December of 2007 when the "Bali roadmap"—a two-year framework for global negotiation adopted at the U.N. climate change convention on the island of Bali—acknowledged that evidence for global warming is "unequivocal" and that delays in reducing greenhouse gas emissions increase the risk of "severe climate change impacts." Although the U.S. continued to show reluctance to accept the economic costs and consequences of cutting emissions, no country in the world could contest any longer that ominous changes in the climate are actually taking place and that coping with them calls for urgent internationally orchestrated action.

FOUR

A Better Way to Grow

Our societies and our economic and financial systems have become unsustainable themselves, and the way they function creates unsustainability in the global ecology. In the course of the past decades, we have been growing the wrong way.

There is a better way to grow. In itself growth is not necessarily bad or even limited: its desirability and future depend on the kind of growth we embark upon. Unrestrained, purely quantitative growth in energy and materials production and consumption is not possible on a finite planet with finite resources and a delicately balanced biosphere: ultimately it is bound to deflect and then turn into growth of a cancerous kind. But there are other forms of growth available to us.

THE CURRENT MODALITY: EXTENSIVE GROWTH

Extensive growth moves along a horizontal plane on the surface of the planet: it conquers ever more territories, colonizes ever more people, and imposes the will of the dominant layers on ever more layers of the population.

The basic end of extensive growth is the extension of human power over larger and larger areas. Traditionally, the means to achieve this end has been conquest: the conquest of nature and the conquest of other, weaker or less power- and domination-oriented peoples. Successful conquest led to the colonization of other tribes, nations, cities, and empires, subjugating them to the ambitions and interests of the conquerors. For most of recorded history this was accomplished by force of arms. Since the second half of the twentieth century it has also been attempted by economic means: wealthy states and global companies using their power to impose their will and values on wide layers of the population.

For states the goal of extensive growth is territorial sovereignty, including sovereignty over the human and natural resources of the territories. The corresponding goal for global companies is to generate demand for consumption, often without regard for the social and environmental consequences.

The ends of extensive growth can be encapsulated in three "Cs": *conquest, colonization,* and *consumption.* These ends are served by corresponding varieties of means: First, the technologies that use and transform matter, the technologies of *production;* second, the technologies that generate the power to operate matter-transforming technologies, *energy-generating* technologies; and third, the technologies that whet people's appetites, create artificial demand, and shift patterns of consumption, the technologies of *propaganda, PR,* and *advertising.* The first of these kinds of technologies built habitations with networks of transportation and communication and increasingly powerful production structures for a growing variety of products. The second harnessed the forces of nature to drive these technologies. The third produced the demand-provoking images and the subtle or not-so-subtle means by which the manufacturers of products and the suppliers of services impose their will on their clients and customers.

A BETTER MODALITY: INTENSIVE GROWTH

The ends of intensive growth are very different from those of extensive growth. Intensive growth centers on the development of individuals and communities. Its principal ends can be grasped under three other "Cs": *connection, communication,* and *consciousness.*

Let us take connection first. One of the great myths of the industrial age has been the skin-enclosed separation of individuals from each other and the disjunction of their interests from the interests of others. The former aspect of this myth has been legitimized by the worldview based on classical physics. Like the mass points of Newton, humans appear to be self-contained, mutually independent chunks of organized matter only externally related to each other and to their environment. Classical economics reinforces this belief by viewing the individual as a self-centered economic actor, pursuing his or her own interests, harmonized at best with the interests of others through the workings of the market. But, as we shall see in part 2, the contemporary sciences no longer support such a view. Now every quantum is known to be intrinsically connected with every other quantum, and every organism with other organisms in the biosphere. In turn, economists know that there is a direct connection between the interests of individuals, individual states, and individual enterprises and the workings of the globalized international system. In our world these embracing connections evolve rapidly, and it is one of the ends of intensive growth to order them, creating coherent structure in place of random proliferation.

The second and third ends of intensive growth are directly linked with the first. These ends are to deepen the level of communication and raise the level of consciousness of the communicators.

Communication unfolds on multiple levels. First of all, we need to communicate with ourselves, caring for and developing our consciousness and personality. People who are "in touch with themselves" are better balanced and more able to communicate with the world around them. We also need to be in communication with those who make up the immediate context of our lives—family, community, and work or profession. Still wider levels of communication are equally necessary: communication with other people, whether near or far, in our own community and in other communities, countries, and cultures.

Communication calls for connection, but on the human plane more enters into play than connection: communication also involves *consciousness*. The full potentials of human communication unfold when the communicators apprehend the strands of connection through which they communicate. A high level of communication calls for a high level of consciousness that enables people to make use of the many, sometimes extremely subtle, strands of connection that bind them to each other and to their environment. Consciousness of these connections could enable us to shift from today's power- and conquest-hungry Logos-civilization to a Holos-civilization centered on the growth of individuals and the sustainability of human communities and the biosphere.

FIVE

A New Vision

It seems to me that a totally different kind of morality and conduct, and an action that springs from the understanding of the whole process of living, have become an urgent necessity in our world of mounting crises and problems. We try to deal with these issues through political and organizational methods, through economic readjustment and various reforms; but none of these things will ever resolve the complex difficulties of human existence, though they may offer temporary relief. . . .

But there is a revolution which is entirely different and which must take place if we are to emerge from the endless series of anxieties, conflicts, and frustrations in which we are caught. This revolution has to begin, not with theory and ideation, which eventually prove worthless, but with a radical transformation in the mind itself.

J. KRISHNAMURTI, "ON LEARNING"

This statement was not written in the beginning of the twenty-first century, but in the middle of the twentieth. There were crises and problems—a "mounting series of anxieties, conflicts, and frustrations"—already in the past century, although only keen intellects were aware of them. A few perceived that these crises can only be coped with by a radical transformation in the mind: a transformation of the values and the beliefs—the vision—we have of ourselves, of others, and of the world around us.

In this opening decade of the twenty-first century the values and beliefs we live by have become sadly, and in some cases dangerously, outdated. We need a new vision of ourselves, our world, and our place and role in the world. The vision we need can be, and indeed must be, grounded in the best insights we have into human nature and the nature of the world in which the human world is embedded.

The vision of the world in which modern people place their trust is the one they consider scientific. This vision is based largely on the physics of Newton, the biology of Darwin, and the psychology of Freud. However, these conceptions have been overtaken by new discoveries. In light of the emerging insights, the universe is not a lifeless, soulless aggregate of inert chunks of matter, rather it resembles a living organism. Life is not a random accident, and the basic drives of the human psyche include far more than the drive for sex and self-gratification. Matter, life, and mind are consistent elements within an overall process of great complexity yet coherent and harmonious design. Space and time are united as the dynamic background of the universe. Matter is vanishing as a fundamental feature of reality, retreating before energy, and continuous fields are replacing discrete particles as the basic elements of an energy-bathed and information-filled universe. The reality we call universe is a seamless whole, evolving over eons of cosmic time and producing conditions where life, and then mind and consciousness, can emerge.

This is a very different vision from that of a mechanistic universe heading inexorably toward the ultimate "heat-death" where nothing new can possibly take place, where all processes have been played out and all free energies have been expended. It is very different from the vision of humans who are separate from each other and are only affecting their immediate environment and where all things can be manipulated the way one manipulates a machine.

Before considering the kind of vision suggested by the latest developments in the sciences, let us look at the kind of vision these developments run counter to: our long cherished beliefs about ourselves, our world, and our responsibilities for each other and for our world.

NINE OUTDATED BELIEFS

1. *Everyone is unique and separate.* We are all unique and separate individuals enclosed by our skin and pursuing our own interests. We have only ourselves to rely on; everyone else is either friend or foe, at best linked to us by temporarily coinciding interests.

2. *Everything is reversible.* The problems we experience are but interludes after which everything goes back to normal. All we need to do is manage the difficulties that crop up using tried and tested methods of problem solving and, if necessary, crisis management. Business as unusual has evolved out of business as usual and sooner or later will reverse back into it.

3. *Order calls for hierarchy.* Order in society can only be achieved by rules and laws and their proper enforcement, and this requires a chain of command that is recognized and obeyed by all. A few people on top (mostly males) make up the rules, legislate the laws, give the orders, and ensure compliance with them. Everyone else is to obey the rules and take his or her place within the social and political order.

4. *Efficiency is the key.* We must get the maximum out of every person, every machine, and every organization, regardless of what is produced and whether or not it serves a humanly and socially useful purpose.

5. *Technology is the answer.* Whatever the problem, technology can already offer the solution—and if not, it can be developed to offer it.

6. *New is always better.* Anything that is new is better than (almost) anything that stems from last year or the year before.

7. *My country, right or wrong.* Come what may, I owe allegiance only to one nation, one flag, and one government.

8. *The more money I have, the happier I am.* There is a direct link between having money and being happy. (A Gallup survey confirmed this belief: three in four young Americans entering college consider it "essential" or "very important" to become very well off financially.)

9. *The future is none of my business.* Why should I worry about the good of the next generation? Every generation has always had to look after itself, and the next generation will have to do the same.

Why these beliefs are obsolete is not difficult to see.

1. That we are unique is true, but it does not mean that we are separate from each other and from nature. Seeing ourselves as separate from the world in which we live distorts our natural impulses to seek our advantage into unfair and unproductive struggles among unequal competitors. Solidarity based on a sense of oneness with others and with nature is a basic condition for creating a world that is peaceful and sustainable.

2. If we remain convinced that the problems we encounter are but temporary interludes in an unchanging and perhaps unchangeable status quo, no experience of the problems will change our thinking and we shall be unable to learn and cope with them.

3. Order is seldom tenable when it is based on hierarchy. Male-dominated hierarchies do not work well even in the army and the church, much less in business and society. Successful business managers have learned the advantages of lean structures and teamwork, but for the most part social and political institutions still operate in the hierarchical mode. As a result, government tends to be heavy-handed, its workings cumbersome and inefficient.

4. Efficiency without regard to what is produced and whom it will benefit is not the answer either. It leads to mounting unemployment, caters to the demands of the rich without regard to the needs of the poor, and polarizes society into "monetized" and "traditional" sectors.

5. Technology is a powerful and sophisticated instrument, but it is only an instrument: its utility depends on what a technology is and how it is used. Even the best technology is a two-edged sword. Nuclear reactors produce an almost unlimited supply of energy, but their waste products and their decommissioning pose serious problems.

Genetic engineering can create virus-resistant and protein-rich plants, improved breeds of animals, vast supplies of animal proteins, and microorganisms capable of producing proteins and hormones and improving photosynthesis, but it can also produce lethal biological weapons and pathogenic microorganisms and destroy the diversity and the balance of nature. In turn, information technologies can create a globally interacting yet locally diverse world, enabling all people to be linked whatever their nationality, culture, and ethnic origin, but information networks dominated by the power groups that brought them into being serve only the interests of that minority and marginalize the rest.

6. That the new is always better is evidently not true. Sometimes the new is worse than what it replaced—more expensive, less enduring, more complex, less manageable, and more damaging to our health and to our environment.

7. The chauvinistic slogan "my country, right or wrong" asks people to fight for causes their government espouses, and may later repudiate, and to embrace the values and worldviews of a small group of political leaders. It ignores the interdependence and the shared future of all people on the globe. There is nothing in the normal human mind that forbids the expansion of one's loyalty above the level of one's country; we are not constrained to swear exclusive allegiance to one flag only. We can be loyal to our community without giving up loyalty to our province, state, or region. We can be loyal to our region and feel at one with an entire culture, and even with the human family as a whole. Americans are New Englanders, Texans, Southerners, and Pacific Northwesterners as well as Americans. Europeans are English, German, French, Spanish, and Italian as well as Europeans. In all parts of the world people can have multiple identities and evolve multiple allegiances.

8. The belief about the link between wealth and happiness is not borne out by experience. Money can buy many things but not happiness and well-being. It can buy sex but not love, attention but not caring, information but not wisdom. Since 1957, the GNP in the United

States has more than doubled, but the average level of happiness has declined: those who report being "very happy" are only 32 percent of the population. At the same time the divorce rate doubled, the teen suicide rate more than doubled, violent crime tripled, and more people than ever say they are depressed. We have big houses and broken homes, high income and low morale, secured rights and diminished civility.

9. Finally, living without conscious forward planning—although it may have been sufficient in days of rapid and seemingly unlimited growth when every new generation appeared able to take care of itself—is not a responsible option at a time when what we do today has a profound impact on the well-being, and even the survival, of our children and our children's children.

SIX PARTICULARLY DANGEROUS MYTHS

1. *Nature is inexhaustible.* The origins of the myth that nature is an infinite source of resources and an infinite sink of wastes go back thousands of years. It would hardly have occurred to the inhabitants of ancient Babylonia, Sumer, Egypt, India, or China that the environment around them could ever be exhausted of the basic necessities of life—edible plants, domestic animals, clean water, and breathable air—or fouled by dumping waste and garbage. The environment appeared far too vast to be much affected by what humans did in their settlements and on the lands that surrounded them.

   Over the course of the centuries, this proved to be a dangerous belief. It turned much of the Fertile Crescent of biblical times into the Middle East of today: a region with vast areas of arid and infertile land. In those days, people could move on, colonizing new lands and exploiting fresh resources. But today there is nowhere left to go. In a globally extended industrial civilization wielding powerful technologies, the belief in the inexhaustibility of nature gives free rein to the overuse and thoughtless impairment of the resources of the planet and the unreflective overload of nature's self-regenerative capacities.

2. *Nature is like a giant mechanism.* This myth dates from the early modern age, a carryover from the Newtonian view of the world, according to which causes have direct and singular effects. The idea of the world as a giant mechanism was well adapted to creating and operating medieval technologies—water mills, windmills, pumps, mechanical clocks, and animal-drawn plows and carriages—but it fails when it comes to living organisms and the world that sustains living organisms. Yet the myth persists that we can engineer the environment as if it were a machine. This creates a plethora of "side effects," such as the degradation of water, air, and soil, the alteration of the climate, and the impairment of local and continental ecosystems. The myth that nature is like a mechanism, although not as old as the myth that it is inexhaustible, is becoming just as dangerous.

3. *Life is a struggle where only the fittest survive.* This myth dates from the nineteenth century, a consequence of the popular understanding of Darwin's theory of natural selection. It claims that in society, as in nature, "the fittest survive," meaning that if we are to survive we have to be fitter for the struggle of life than others around us: smarter, more ambitious, more daring, and richer and more powerful than our competitors.

   Transposing nineteenth-century Darwinism to the sphere of society can be lethal, as the "social Darwinism" adopted by Hitler's Nazi ideology has shown. It justified the conquest of territories and the subjugation of other peoples in the name of racial fitness and purity. In our day, the varieties of social Darwinism include but also go beyond armed aggression to the more subtle but in some ways equally merciless struggle of competitors in the marketplace. No-holds-barred competition produces widening gaps between rich and poor and concentrates wealth and power in the hands of a shrinking minority of unscrupulous managers and speculators. It relegates states and entire populations to the role of clients and consumers, and if poor, dismisses them as marginal factors in the equations that determine success in the marketplace.

4. *The market distributes benefits.* The myth of the market is directly related to the survival-of-the-fittest myth and is often cited as justification for it. Unlike in nature, where the consequence of "fitness" is the spread and dominance of a species and the extinction or marginalization of others, the market myth tells us that in society there is a mechanism that distributes the benefits instead of having them accrue only to the fit. This is the free market, governed by what Adam Smith called the "invisible hand." It acts equitably: if I do well for myself, I benefit not only myself, my family, and my company, but also my community. Wealth "trickles down" from the rich to the poor: a rising tide lifts all boats.

   The market myth is comforting for the rich, but it disregards the fact that the free market distributes benefits only under conditions of near-perfect competition, where the playing field is level and the players have a more or less equal number of chips. In the real world, the playing field is not level and the distribution of wealth is strongly skewed. Not surprisingly, in today's world the poorest 40 percent is left with 3 percent of the global wealth and the wealth of a few hundred billionaires equals the annual income of three billion of the world's poor people.

5. *The more you consume the better you are.* According to this myth there is a strict equivalence between the size of your wallet and your personal worth as the owner of the wallet. The equating of human worth with financial worth has been consciously fueled by business; companies did not hesitate to advertise unlimited consumption as a realistic possibility and conspicuous consumption as the ideal. Fifty years ago retailing analyst Victor Lebow gave a clear formulation of the consumption myth. In his book *How Much Is Enough?* Alan Durning quotes: "Our enormously productive economy demands that we make consumption our way of life, that we convert the buying and use of goods into rituals, that we seek our spiritual satisfaction, our ego satisfaction, in consumption. The economy needs things consumed, burned, worn out, replaced, and discarded at an ever-increasing rate."

The consumption myth remains powerful today, even if it is not as brazenly stated as before. In constant dollars the modern world has consumed as many goods and services since 1950 as in all previous generations put together—and with China and India entering the consumption spree, it will consume as much again in much less time.

6. *Economic ends justify military means.* The ancient Romans had a saying: "If you aspire to peace, prepare for war." For them this made sense: the Romans governed a global empire, with rebellious peoples and cultures within and barbarian tribes at the periphery. Maintaining it required a constant exercise of military power. Today the nature of power is different, but the belief about the use of war to achieve political—and now also economic—objectives is much the same. Like the ancient Romans, during the Bush administrations the U.S. believed that maintaining world supremacy called for "sending in the marines." But the twenty-first century world is not the classical world: it is more interactive and interdependent, and its social, economic, and ecological systems operate dangerously close to the edge of sustainability. In this world the belief that war is the way to achieve economic and political objectives is a myth and—in view of its human, social, and even ecological consequences—it is a dangerously obsolete myth.

SHEDDING OUTDATED MYTHS AND BELIEFS

How does one shed obsolete myths and beliefs—how does one make the leap to a vision that embraces the human being in his and her planetary environment? In contemporary societies a number of factors hinder the shift to a better vision. The way children are raised depresses their faculties for learning and creativity; the way young people experience the struggle for material survival results in frustration and resentment. In adults this leads to a variety of compensatory, addictive, and compulsive behaviors. The result is the persistence of social and political oppression, merciless competition for resources and markets, cultural intolerance, crime, and disregard for the environment. To live with each other and not against each other, to live in a way that does not rob others of the chance to live, to care what is happening to the poor and the powerless as well as to nature, all this calls for a more mature outlook—a significant measure of inner growth.

In today's world achieving true inner growth is not easy, yet some people have achieved it nevertheless. Some among them have had life-transforming experiences. Astronauts, for example, had the privilege of looking at Earth from outer space; they have seen a precious world without boundaries, a home to all humans and all living things. They came back different people. They realized how petty and superficial it is to squabble over privileges and powers when we live on a resplendent planet unique in this corner of the universe.

Another life- and mind-transforming experience is the experience of coming back from the portals of death. People who have had a near-death experience return to everyday life with a deeply altered consciousness. They no longer fear dying. They achieve inner peace, have empathy for others and reverence for nature, and have a fresh appreciation of the wonder of existence.

Deep religious and spiritual experience is also conducive to inner growth. People who engage in intense meditation or prayer know that differences among people, whether due to sex, race, color, language, political conviction, or religious belief, do not mean that they are separate from each other. They recognize that William James was right: we are like islands in the sea, separate on the surface but connected in the deep. There are levels of existence through which we not only communicate with each other—we become part of each other.

There are more accessible paths to inner growth as well. We can train ourselves to achieve greater unity between our bodies and minds. Most of us have lost contact with our bodies. We are constantly occupied and preoccupied with tasks and aspirations, with hopes, fears, and worries. We use our bodies as we use our cars or computers: giving them commands to take us where we want to go and do what we want to have done. We live in our heads, with little time and inclination to live in our whole bodies. We are losing the ground under our feet.

Grounding ourselves in our bodies is a first step; it needs to be followed by another. The stresses and strains of existence also impact our emotional lives, and those too need attention. It is not that we have lost contact with our emotions—we are only too aware of them much of the time. But they are often the wrong kind of emotions. Negative feelings, such as anger, hate, fear, anxiety, suspicion, jealousy, contempt, and indifference, dominate the tenor of life in modern societies. They result from lifetime experiences that are mainly negative. With some exceptions, even childhood education is based on negative reinforcements such as punishment and the threat of failure. Positive emotions of love and caring are the preserve of the family and our circle of friends, but these aspects of life are often sacrificed to the pressure of work and the struggle to secure our livelihood. Positive emotions can be created in the context of loving and caring relations with those around us and can also be generated by experiences of nature: beholding the tranquillity of a sunny meadow or a calm lake, the beauty of a sunset, the majesty of a mountain, or the awesomeness of a stormy sea.

THE TEN COMMANDMENTS OF A TIMELY VISION

The vision we need to live on this planet without destroying ourselves and its delicate web of life can be formulated in reference to ten "commandments."

1. Live in ways that enable other people to also live, satisfying your needs without detracting from the chances of others to satisfy theirs.
2. Live in ways that respect the right to life and to economic and cultural development of all people, wherever they live and whatever their ethnic origin, sex, citizenship, station in life, and belief system.
3. Live in ways that safeguard the intrinsic right to life and to an environment supportive of life of all the things that live and grow on Earth.

4. Pursue happiness, freedom, and personal fulfillment in harmony with the integrity of nature and with consideration for the similar pursuits of others.

5. Require that your government relates to other nations and peoples peacefully and in a spirit of cooperation, recognizing the legitimate aspirations for a better life and a healthy environment of all the people in the human family.

6. Require of the enterprises with which you do business that they accept responsibility for all their stakeholders as well as for the environment, and demand that they produce goods and offer services that satisfy legitimate demand without reducing the chances of smaller and less privileged entrants to compete in the marketplace.

7. Require of the public media that they provide a constant stream of reliable information on basic trends and crucial processes to enable you to reach informed decisions on issues that affect your life and well-being.

8. Make room in your life to help those less privileged than you to live a life of dignity, free from the struggles and humiliations of abject poverty.

9. Encourage young people and open-minded people of all ages to evolve the spirit that could empower them to make ethical decisions of their own on issues that decide their future and the future of their children.

10. Work with like-minded people to preserve or restore the essential balances of the environment, with due attention to your neighborhood, your country and region, and the whole of the biosphere.

SIX

A Planetary Ethic

We have arrived at a historic bifurcation, at the critical phase of the global Macroshift. While we now find ourselves on a descending path toward growing social, political, and environmental crises, we could also enter on an ascending path leading to a system of social, economic, and political organization that is peaceful and capable of ensuring sustainability for human communities and the planetary environment.

The choice is open, and making it depends on our values, beliefs, and vision and—as we shall see—on our ethic.

In a globally interacting and interdependent world the dominant ethic cannot be local, regional, or ethnic. It cannot be only a Christian ethic, or Jewish, Muslim, or Hindu ethic—it must bring together all the ethics that guide people's behavior: what is needed is a *planetary ethic*.

The principles of an effective planetary ethic need to be translated into codes that define the kind of behavior all people can agree is moral and desirable. Finding such codes is in everybody's best interest.

In the past, creating and disseminating moral codes was the task and the privilege of the great religions. Examples of such codes are the Ten Commandments of Jews and Christians and the Buddhist Rules of Right Action. But today the power of religious doctrine–based codes for moral behavior has been diminished by the advance of science. Yet, even if science has displaced religion as a source of authority, scientists have not come up with alternative moral codes. There have been a few attempts, but they were abandoned. Henri de Saint-Simon in the late 1700s, Auguste Comte in the early 1800s, and Emile Durkheim in the late 1800s and early 1900s were at pains to develop a set of "positive"—that is, scientific observation- and experiment-based—codes for public morality. This endeavor was so contrary to the underlying "pure objectivity" ideals of twentieth-century science that it was not taken up by mainstream scientists and philosophers.

In the 1990s, however, both scientists and political leaders began to recognize the need for principles that would state universal norms for behavior. In April 1990, in the "Universal Declaration of Human Responsibilities," the InterAction Council, a group of twenty-four former heads of state or government declared, "Because global interdependence demands that we must live with each other in harmony, human beings need rules and constraints. Ethics are the minimum standards that make a collective life possible. Without ethics and the self-restraint that are their result, humankind would revert to the survival of the fittest. The world is in need of an ethical base on which to stand."

The Union of Concerned Scientists, an organization of leading scientists, concurred. "A new ethic is required," claimed a statement signed in 1993 by 1,670 scientists, including 102 Nobel laureates, from 70 countries. "This ethic must motivate a great movement, convincing reluctant leaders and reluctant governments and reluctant peoples themselves to effect the needed changes." The scientists noted our new responsibility for caring for the Earth and warned that "a great change in our stewardship of the Earth and the life on it is required if vast human misery is to be avoided and our global home on this planet is not to be irretrievably mutilated." Human beings and the natural world, they said, are on a collision course, one that could so alter the living world that it will be unable to sustain life as we know it.

In November 2003, a group of Nobel Peace laureates meeting in Rome affirmed, "Ethics in the relations between nations and in government policies is of paramount importance. Nations must treat other nations as they wish to be treated. The most powerful nations must remember that as they do, so shall others do." And in November 2004, the same group of laureates declared, "Only by reaffirming our shared ethical values—respect for human rights and fundamental freedoms— and by observing democratic principles, within and amongst countries, can terrorism be defeated. We must address the root causes of terrorism—poverty, ignorance and injustice—rather than responding to violence with violence."

To convince modern people to adopt behaviors appropriate to the conditions that now reign on Earth—something that many people are reluctant to do—there must be meaningful codes to guide people's behavior: *all* people's behavior. Such codes need to be distilled from a planetary ethic appropriate to the interdependence and shared destiny of the human community.

ECOLOGICAL ETHICS: THE FIRST STEP TOWARD A PLANETARY ETHIC

In the field of ethical theory the closest development to a planetary ethic is the branch of environmental ethics known as ecological ethics. Ecological ethics aims at harmonizing the rhythms, dynamics, and conscious or unwitting effects of human life on nature with the rhythms and dynamics of nature. At its best, an ecological ethic is an ethic of sustainable human impact on the biosphere.

The field of ecological ethics is a newcomer in the spectrum of ethical theories. For virtually the entire duration of Western intellectual history ethical discussion failed to manifest a direct concern with obligations that humans might have toward the natural environment. This unconcern extended even to other living things, such as plants and animals. Their value was seen mainly in reference to the values entertained by humans. At the most, plants and animals were seen to have an indirect, *instrumental* value, according to whether they contributed to, or detracted from, the realization of humanly held values. Ethical commitments and moral obligations rested on a human person-to-person basis: people as individuals have ethical commitments to family, friends, and fellow-citizens; together they constitute a moral community with acknowledged obligations to each other.

Ethical commitment has then been extended to communities other than one's own. Moral philosophers consider that a moral community has ethical commitments not only to its own members but to communities throughout the continents. Ultimately, with the advent of the debate on our responsibility to future generations, ethical commitments have also been extended to human communities across time as well as space.

But the environment of human communities was not included in the range of ethical commitments. At the most, certain elements of it were seen to have a derivative value, inasmuch as they contributed to the health, well-being, or social and spiritual fulfillment of human beings, the sole subjects of ethical discourse. For Western common sense, such an instrumental approach to nature made eminent sense. The carrier of value, it was said, is the human being, for he or she alone has a conscious mind where values can be entertained, and hence satisfied or frustrated.

Instrumental value approaches to the environment dominated up to the 1970s. Until then the questions philosophers posed in regard to other species mainly concerned their relevance to humans and human interests. This attitude extended even to endangered species. The rose-colored periwinkle, for example, is a plant in danger of extinction on Madagascar but, as Timothy Weiskel noted, the morality of this situation has been typically assessed in terms of the periwinkle's use as a source of a natural anticarcinogen.

There is also another kind of value, however, known as *intrinsic* value. Traditionally, intrinsic values have been assigned only to humans: human life, as Immanuel Kant made clear, must always be an end in itself and never a means to something else. Humans clearly have intrinsic value, and in the 1980s the question arose whether intrinsic value must be limited to humans. A few thinkers, among them the Norwegian philosopher Arne Naess, insisted that not only our fellow humans but also other forms of life around us must be valued in and for themselves. With the advent of what Naess called deep ecology, instrumental value approaches to non-human nature were replaced by the attribution of intrinsic value to all things in the biosphere.

REVERENCE FOR NATURAL SYSTEMS

In a planetary ethic Albert Schweitzer's famous tenet, *reverence for life*, serves as a basis for attributing intrinsic values to things around us. Life, as scientists know, is not an entirely separate category: it is a set of phenomena—such as metabolism and reproduction—that appear at a certain point in the evolution of complexity. The basic process was the evolution of physical and physical-chemical systems into bacteria and primitive, so-called prokaryotic cells that subsequently evolved into more evolved eukaryotic cells. Unicellular systems first formed colonial organisms where some level of specialization among them created a collective system, manifesting some of the phenomena we associate with life, and later formed genuine, multicellular organisms of which the many species and varieties populate the biosphere. In this process there is no place where we can draw a line between life and non-life. Some systems still straddle the divide; viruses, for example, appear to be physical-chemical nonliving systems when separated from a host and living systems when associated with a host.

Thus the kind of system to which a planetary ethic can attribute intrinsic value is not just the living system but the general category of system in which the living system arises. The simplest denomination of this general category is "natural system."

A natural system can be defined as that variety of system (initially physical-chemical, then—higher in the scale of evolution—biological, and then sociocultural) that arises independently of conscious human design and its implementation. In a planetary ethic reverence for this variety of system is expressed in the attribution of intrinsic value: value for such systems in themselves, regardless of their relation to and utility for other systems.

A natural system is a broad but not a universal category; if all systems in the world were natural systems, the definition would lose meaning. It is true, of course, that in a deep spiritual sense intrinsic value can be attributed to all things in space and time—to the universe in its totality. But this all-inclusive valuation fails when it comes to deriving practical criteria for moral behavior. There must be things to which we attribute intrinsic value, and other things that we value instrumentally, and still others to which we do not attribute specific value (although we should recognize that all things are part of the wholeness of the universe and are therefore "holy" in a spiritual sense).

We can distinguish systems that are not natural systems in regard to their relationship to natural systems. Among the non-natural systems, some are directly relevant to the existence and evolution of natural systems, while others are only indirectly and insignificantly relevant. The former kind of nonnatural system is an element in the relevant environment of natural systems and is instrumental in furthering, or else hindering, their existence and evolution. It is logical to place instrumental value on systems that form part of the relevant environment of natural systems.

Thus, in a planetary ethic we attribute *intrinsic value* to natural systems (that is, to systems that arise, subsist, and evolve basically independently of conscious human planning and execution) and *instrumental value* to systems that are relevant to the existence of natural systems.

How can we delimit the category of natural system so we can distinguish such systems from other kinds of systems? Clearly, the universe in its totality is a natural system, for it arises, subsists, and evolves independently of human design and implementation. However, identifying the universe in its totality as a natural system, as just noted, is too inclusive; it needs to be further specified.

A first delimitation of the concept of the universe as a natural system is the consideration that our planet's biosphere is such a system. Physicists and ecologists point out that the entire biosphere is a thermodynamically open system, fed with free energy from the Sun. Were the energy differential between the surface of the Sun (approximately 11,000 degrees F) and the surface of the Earth (about 77 degrees F) ever to equalize, only worms and clams at the bottom of the deepest oceans could survive for any appreciable amount of time. Plants use sunlight in photosynthesis, converting water and carbon dioxide into carbohydrates; animals eat plants and other animals; and humans at the top of the food chain eat both plants and animals. As figure 2 illustrated, the solar irradiation "drives" this thermodynamic system, enabling it to access, store, and utilize ever more of the incoming energy. As a result the web of life on Earth has distinct systemic properties. All elements are related to all others, and changes in one propagate at some level throughout the whole system.

The biosphere as a whole is a natural system, but without further specification it is still too vast to permit the derivation of practical codes of moral behavior. We can, however, make further distinctions as regard sub-biospheric systems. Some of these systems are natural systems, and others are not. Those entities that do not owe their existence to human volition are natural systems. This set includes all living organisms. It includes human beings as individual organisms.

There is a question, however, whether the category of natural system also includes humans *collectively*. After all, groups of humans have conscious, volitional members, and they form identifiable systems of their own. Are these systems the result of the conscious, volitional, and perhaps arbitrary actions of their members? If they are, they must be excluded from the set to which a planetary ethic can attribute intrinsic value.

Yet, on deeper analysis, human groups of a certain kind do form natural systems. Despite the fact that their members are individually conscious beings, historically formed communities are not the outcome of conscious volitional acts. They are not intentionally organized systems, even if they are the joint outcome of a large number of intentional acts and behaviors. Historical communities subsist, evolve, and degenerate largely independently of the conscious intentional designs and activities of their human members, even if they are shaped—and on occasion strongly shaped—by their members' designs and activities.

In a planetary ethic individual human beings as well as the communities formed by human beings qualify for intrinsic value. This does not mean that all things in nature would do so. There is a multitude of atoms and molecules that are not integrated in the functional structure of living systems and of the biosphere: they make up the atmosphere and the oceans, rivers, lakes, soils, mountains, and deserts of the globe. Together with the Sun—which provides the free energy that drives the planet's biological and ecological processes—they constitute the physical environment of living systems. These entities qualify for the attribution of *instrumental* value.

The next and last question we must ask is whether there are systems that are neither natural systems nor relevant to the existence and evolution of natural systems. The answer is that there are systems of this kind in interstellar space and deep under the layers of the Earth's crust. But, given that the biosphere is a natural system interacting with its physical environment, there are no such entities on the surface of the Earth.

Let us summarize. A planetary ethic attributes *intrinsic* value to the web of life that has evolved on this planet and also to the things, from algae to ecologies and human beings and societies, that have emerged in that web and form a part of it. It attributes *instrumental* value to the physical environment of the web of life, including the atmosphere, the hydrosphere, and the geosphere, since it provides resources and physical and chemical conditions for the subsistence and evolution of natural systems.

THE DERIVATION OF PRACTICAL BEHAVIORAL CODES

The above concepts appear abstract and theoretical, yet they have application in practice: a set of meaningful moral codes can be derived from them. These codes respond to specific criteria. They specify behavior consistent with the intrinsic valuation of natural systems and the instrumental valuation of systems in their relevant environment.

Two kinds of codes need to be distinguished. One is the *maximum code*: it commits individuals to positive ends that contribute to the existence and evolution of natural systems in the biosphere, above all, to human beings and their communities. The other is the *minimum code:* it requires that, at the least, individuals limit the negative impact of their behavior on natural systems.

The Maximum Code

In a first approximation the maximum code can be formulated as follows: *Act so as to maximize the sustained persistence of the biosphere.* This code requires that we take care of nature and adapt ourselves to its patterns and refrain from reducing biodiversity, disturbing natural balances, and modifying or destroying vital energy and information flows.

However, a sound planetary ethic must adopt a dynamic rather than a static view of the biosphere. More is involved than maximizing the sustainability of the systems of nature as they are presently constituted. In the dynamic view, the biosphere and its systems undergo structural and functional changes, some piecemeal and linear, others more radical and nonlinear. To remain valid and effective, moral behavior must be adapted not to the current state but to the foreseeable *evolution* of the intrinsically valued natural systems. We should rephrase the maximum code to state: *Act so as to further the ongoing evolution of the biosphere.*

On a second look, even the modified maximum code turns out to be flawed. This is because biological and sociocultural-technological evolution on this planet has an unsettling feature: it is constantly accelerating. Not only have more and more forms of life evolved in the course of the last 3.5 billion years, but the rate at which they evolve has also been speeding up. Over half of evolutionary time was taken up with the advance from the stage of nonnucleated prokaryotes to that of nucleated eukaryotes; thereafter it took evolution half that time to reach the level of fish. Since then the time intervals between the major evolutionary steps have constantly shortened. The Miocene epoch is about 25 million years in the past, the Lower Pleistocene of the Quaternary began 1.6 million years before our time, the Middle 750,000 years, and the Upper merely 125,000 years. Hominid creatures appeared during the Holocene (or Recent) epoch, though our lineage may have diverged from other hominoid species much before then.

Thereafter human evolution shifted from the genetic to the socio-cultural realm. Organized societies with rites of passage, writing, and other sociocultural practices appeared about 20,000 years ago; the first varieties of plants and animals were domesticated eight to ten thousand years before our time; and the great empires of the Middle and Far East appeared a few thousand years later.

At the dawn of the modern age the evolution of human societies was accelerated by powerful technologies. By the twentieth century, driven by newly discovered sources of energy, first steam and then coal, oil, and natural gas, and impelled by other technologies for process-ing, storing, and transmitting information, the rate of societal evolution became vertiginous. It is difficult to see how humans, with a physiology governed by a brain and nervous system that evolved a hundred thou-sand or more years in the past, could keep pace with a continuation of this trend. Furthering it, even in the modest sense of making it relatively smooth, is likely to be at best a temporary measure. Sooner or later the processes of societal change will accelerate beyond the capacity of flesh-and-bone humans to keep up. A range of ecological catastrophes could be triggered, and ultimately our species would join the ranks of the over 99 percent of vertebrates that have become extinct since the Cambrian period.

This further reflection leads to the conclusion that we need to do more than further the evolutionary process. The indicated maximum code must aim at *stabilizing* the process at a humanly favorable level. Thus the correct formulation of the code is: *Act so as to further the evo-lution of a humanly favorable dynamic equilibrium in the biosphere.*

This formulation of the code may appear unduly anthropocentric. Instead of respecting the evolution of the biosphere and its natural sys-tems, it aims to intervene in the human interest. Is this not a restate-ment of the classical, and now largely discredited, Baconian approach to nature?

The above formulation is indisputably anthropocentric, but not unduly so. It rests on the consideration that as a species we have the natu-ral capability, and hence the natural right, to exercise our drive for collective survival. Doing so is defensible as long as it does not interfere with the similar capability—and right—of other species. Respecting the maximum code would largely (if not completely) satisfy this proviso. Interventions that seek to establish a form of dynamic balance in the biosphere that is hospitable to humans and favorable to their well-being are likely to involve safeguarding most (though perhaps not all) the species and ecologies that currently exist. In the final count what is good for humans is good—with at the most a very few exceptions—for all life on Earth.

The Minimum Code

Ensuring the dynamic equilibrium of the biosphere at a humanly favorable level requires a thorough restructuring of human relations to the environment, transforming them from the currently unsustainable to a long-term sustainable mode. It also requires the transformation of the institutions of human societies so they can coexist with each other and with nature productively and in peace. These transformations call for a fundamental change in the behavior of individuals, nations, and businesses. Consequently the maximum code, though a long-term goal and an ideal, is utopian in the short-term as a guide to action. It needs to be complemented with a behavioral code that is more immediately practicable. This calls for adding a "floor" to the maximum code "ceiling." This floor is the minimum code for acceptable moral behavior: the *sine qua non* of action that can qualify as moral.

The minimum code updates the classical laissez-faire tenet, "Live and let live." The classical tenet is outdated: On an interdependent and increasingly crowded and resource-depleted planet, letting people live in any way they may wish is not morally permissible. The rich and the mighty would (as they already do) consume a disproportionate share of the planet's resources and voluntarily or inadvertently block access to vital resources by the less privileged. Thus we need a minimum code that asks people to respect the conditions under which all people can live. The indicated minimum code is: *Live so that others can also live.*

The above code derives from Kant's categorical imperative: "Act so as to allow your action to become a universal maxim." In the context of a planetary ethic the minimum code specifies that it is imperative that all people act in a way that can be replicated by all other people without destroying the vital balances of the biosphere or impelling its evolution beyond the threshold of a humanly favorable dynamic equilibrium.

Respecting even the minimum code for individual and collective behavior calls for a series of changes and adaptations. The privileged strata of society still live in a way that the less privileged strata could not duplicate, for the planet has neither the resources nor the carrying capacity for all people to drive private cars, live in separate homes, and use the myriad gadgets and appliances that go with the lifestyle of the affluent. Changes are also called for on the part of the less privileged: they, in turn, must cease to emulate—or try to emulate—the lifestyles of the rich. It would not be enough for Americans, Europeans, Japanese, and Australians to reduce harmful emissions and economize on energy if the Chinese people and those of the rest of the developing world acquired Western driving and consumer habits and continued to burn coal for electricity and wood for cooking.

In the fifth century BCE, in the *Tao Te Ching*, Lao-tzu wrote,

> One's individual life serves as an example for other individuals; one's family serves as a model for other families; one's community serves as a standard for other communities; one's state serves as a measure for other states; and one's country serves as an ideal for all countries.

The minimum code is to ensure that the example set by one individual is worthy of becoming a standard, a measure, and perhaps an ideal, for all others.

IN SUMMARY

If accepted and adopted, the maximum and the minimum codes for moral behavior would ensure a reasonable chance of achieving a sustainable and humanly favorable dynamic equilibrium in the biosphere.

The minimum code would defuse resentment and animosity arising from uneven levels of economic development; it would reduce the potential for conflict based on inequalities in living standards and access to resources. The maximum code, in turn, would create an impetus to move purposefully to the next plateau of dynamic equilibrium between human societies and their life-supporting environment.

In the final analysis the minimum code would create breathing space, buying time for the necessary behavioral changes, while the maximum code would offer an ideal toward which to strive when the time is ripe for such changes.

SEVEN

The Culture of Holos

Moving toward a civilization of Holos is not merely an option: it is a survival imperative. Fortunately it is not unfeasible, nor is it unprecedented. The kind of shift it entails is part of the evolution of human societies, an evolution that began with the mythic civilizations of the Stone Age, continued with the theocratic civilizations of the archaic empires, and moved to the human reason-based civilizations initiated by the ancient Greeks. Now the reign of Logos is drawing to a close: the short-term rationality underlying the currently dominant form of civilization produces more heat than light—more negative social, economic, and ecological consequences than positive, humanly desirable outcomes.

The time has come for a further shift: from a civilization of Logos to a civilization of Holos.

Reaching a civilization of Holos means a transformation that is not unique in history, but faster than any transformation in the past. Because of the speed with which today's global Macroshift is evolving, many people have not caught up with it: to them a Holos civilization appears utopian. Yet there are others for whom a holistic culture is already the norm. And these people are more numerous than we may think.

THE RISE OF A NEW CULTURE

In many societies an intensely hopeful culture is surfacing. It is made up of people who are rethinking their preferences, priorities, values, and behaviors, shifting from consumption based on quantity toward selectivity in view of quality defined by environmental friendliness, sustainability, and the ethics of production and use. Lifestyles hallmarked by matter- and energy-wasteful ostentation are shifting toward modes of living marked by voluntary simplicity and the search for a new morality and harmony with nature.

These changes in values and behaviors, though generally dismissed or underestimated, are both rapid and revolutionary. They are occurring in all segments of society, but most intensely at the margins. Here a number of grassroots movements are opting out of the mainstream and are reforming themselves. These groups are barely visible since for the most part their members go about their business without trying to convert others or call attention to themselves. They underestimate their own numbers and lack social cohesion and political organization. Yet the more serious and sincere of these emerging cultures merit recognition. Unlike in esoteric sects, members of these cultures do not engage in antisocial activities, indulge in promiscuous sex, or seek isolation. Rather, they try to rethink accepted beliefs, values, and lifeways and to strike out on new paths of personal and social behavior.

Such shifts in the culture of a growing number of people must be taken seriously. Dismissing and distrusting all people who do not accept the current system of values and the associated worldviews and lifestyles is naive and indiscriminate. It is true that some alternative cultures are escapist, introverted, and narcissistic, but the more serious have a genuine core of values and priorities that is highly promising for a positive outcome of the Macroshift. Dismissing them would be throwing out the baby with the bathwater.

An Emerging Culture in the United States

In the United States, at the center of the industrialized world, a hopeful subculture is in rapid growth. This is the surprising conclusion of a series of opinion surveys carried out by organizations and individuals keen on tracing the evolution of the thought and action of Americans.

California's Institute of Noetic Sciences found that the changes that occur in America's hopeful subculture include the following shifts in values and behaviors:

- the shift from competition to reconciliation and partnership
- the shift from greed and scarcity to sufficiency and caring
- the shift from outer to inner authority (from reliance on outer sources of "authority" to inner sources of "knowing")
- the shift from mechanistic to living systems (from concepts of the world modeled on mechanistic systems to perspectives and approaches rooted in the principles that inform the realms of life)
- perhaps most significant of all, the shift from separation to wholeness—a fresh recognition of the wholeness and inter-connectedness of all aspects of life and reality

An important shift is occurring in the area of consumer behavior. In her book *Megatrends 2010* Patricia Aburdene traced the rise of what she calls "conscious capitalism," a trend that appears in the market as conscious, or values-driven, consumption. By the turn of the century the market in the United States for values-driven commerce had reached $230 billion (*The New York Times* called it "the biggest market you have never heard of"). According to Aburdene, conscious consumers—often referred to as LOHAS (Lifestyles of Health and Sustainability) consumers—make up a rapidly growing segment in five sectors of the economy:

- the sustainability sector, including ecologically sound construction, renewable energy technologies, and socially responsible investments
- the healthy living sector, appearing in the market as demand for natural and organic foods, nutritional supplements, and personal care

- the alternative healthcare sector, comprised of wellness centers and complementary and alternative medical services and health care
- the personal development sector, made up of seminars, courses, and shared experiences in the body-mind-spirit area
- the ecological lifestyle sector, appearing in the form of demand for ecologically produced, recycled, or recyclable products, as well as ecotourism

The Fund for Global Awakening implemented a survey aimed at elucidating the shared values and beliefs held by people from diverse backgrounds. Carried out in the framework of the In Our Own Words 2000 Research Program, the survey distinguished eight "American types." It found that 14.4 percent of the 1600 respondents—selected so as to represent a cross section of American society—are centered in a material world, whereas 14.2 percent are disengaged from social concerns, 12.1 percent embrace traditional values, and 10 percent are cautious and conservative. These make up half of the U.S. population: the conservative, traditional half. Another 11.9 percent seeks to connect to others through self-exploration, 9.4 percent persists through adversity, 11.6 percent seeks community transformation, and 16.4 percent works for what the survey defines as a "new life of wholeness." These make up the more creative and at least in part change-oriented half. Among them those who seek community transformation and work for a new life of wholeness make up 28 percent of the people. This segment manifests the values, the vision, and the beliefs that could shift U.S. society toward a holistic civilization.

The above findings match the results of a survey carried out in the late 1990s by public opinion researcher Paul Ray. Ray called the significantly forward-looking and open segment of American society the "cultural creatives." In his surveys this segment contrasts with another subculture in America: the "traditionals," who opt out of the mainstream by harking back to the seemingly ideal conditions of bygone times. They make up 24.5 percent of the U.S. population: 48 million adults, coming from a variety of socioeconomic and ethnic backgrounds, with family incomes in the relatively low range of $20,000 to $30,000 per year, due among other things to the diminished income of the many retirees among them.

The "moderns" represent the mainstream culture of Americans. They are stalwart supporters of consumer society; their culture is that of the office towers and factories of big business and of the banks and stock markets. Their values are those taught in the most prestigious schools and colleges of America. In 1999 this was the culture of some 48 percent of the American people: 93 million out of about 193 million adults, more men than women. Family income was $40,000 to $50,000 per year, situating moderns in the middle to upper income bracket.

In the United States, the alternative culture Ray calls "cultural creatives" is the most hopeful segment of the population. It is made up of people from the middle to the wealthy classes, numbering nearly twice as many women as men. According to Ray, at the turn of the century the share of this subculture was 23.4 percent of the U.S. adult population, slightly less than the 28 percent found subsequently by the *IOOW* survey. The factor that identifies the cultural creatives is less what they preach than what they practice, for they seldom attempt to convert others, preferring to be concerned with their own personal growth. Their behavior, especially their lifestyle choices, differentiate them from the other cultures.

The common thread among the members of this emerging subculture is their holism. This comes to the fore in their preference for natural whole foods, holistic health care, holistic inner experience, whole system information, and holistic balance between work and play and consumption and inner growth. They view themselves as synthesizers and healers, not just on the personal level but also on the community and the national levels, even on the planetary level. They aspire to create change in personal values and public behaviors that could shift the dominant culture beyond the fragmented and mechanistic world of the moderns.

Twenty years ago the cultural creatives made up less than 3 percent of the total but at the turn of the century they totaled over fifty million people—and their numbers are growing.

These trends are not generally known, even by those who participate in them. Moderns firmly believe they are the representative majority and will remain so. Traditionals claim that they are the up-and-coming mainstream, citing as evidence the proliferation of conservative radio stations and the swelling membership of some mega-churches and conservative denominations. Cultural creatives, on the other hand, underestimate their own numbers. Many of them believe that they are no more than 5 percent, or at most 10 percent, of the U.S. adult population—far from their estimated 24 percent or 28 percent share.

The U.S. cultural creatives are not alone: similar subcultures are emerging in many parts of the world. A survey carried out in late 2005 by the Italian branch of the Club of Budapest found that about 35 percent of Italians live and act as cultural creatives. Similar figures are coming to light in surveys in other countries of Europe, as well as in Japan, Australia, and Brazil.

In an article in the May/June 2005 issue of the British journal *Resurgence,* William Bloom, head of the UK-based Holistic Network, noted, "The holistic approach is rapidly becoming a major cultural force. There is substantial and rigorously researched evidence that the majority of the population in the United Kingdom, and other industrialized and democratized nations, is adopting a holistic worldview." Holism, he said, catches the most profound spiritual instincts: to become a fulfilled and whole human being; to create healthy and whole communities, local and global; to include and care for all elements and dimensions; to connect with and feel that we are part of the whole meaning and mystery of existence.

When members of the emerging holistically thinking and acting culture awaken to the fact that they are more numerous and widespread than they think, they will get organized. They could then achieve the kind of social, economic, and political weight that could make them into a major agent of change—a main driver of the shift from a civilization of Logos to a civilization of Holos.

EIGHT

Evolution, Not Extinction!
A CALL FROM FIJI

Adi Da, a New York–born world-renowned hermit living on a remote island of Fiji, has issued an urgent call for change and transformation in the human world. He has recognized the threat of extinction and is calling for evolution. He asks that people come together to create a world united in its determination to achieve peace and sustainability and discover the unity that is basic to all things in the cosmos.*

Every once in a while a prophetic voice is raised in the midst of crisis and chaos. It cuts through the walls of indifference, neglect, and just plain ignorance and exposes the heart of the issue. Adi Da expresses such a voice. Not surprisingly, it comes from one who is not part of the hustle and bustle we charitably call the business of living, and less charitably the daily rat race. It comes from one who decided early in life to keep the distance needed for clear vision and enter the silence needed for true audition. We see things best when we have them in perspective: then we see the forest and not only the trees. And we hear best when we silence

---

*This chapter has been adapted from the author's introduction to Adi Da's book, *Not-Two Is Peace: The Ordinary People's Way of Global Cooperative Order* (Middletown, Calif.: LS Peace 723, 2007).

the cacophony of competing voices clamoring for attention. The source of deep insight is the emptiness that is also a fullness and the profound silence that allows the voice of reason to be heard.

The heart of the issue that Adi Da addresses is none other than the issue of our collective survival—the survival of the species that calls itself *Homo sapiens:* "*Homo* the knower," "*Homo* the wise." We have reached the very edge of our species' viability on this planet. The problems are becoming more evident every day. Adi Da states them succinctly: ". . . environmental pollution, global warming, climate change, the abuse of power by corporations and governments, the necessity for new technologies and new methods in every area of human life, the scarcity of fuel resources and of natural and human resources altogether, disease, famine, poverty, overpopulation, urbanization, globalization, human migration, territorial disputes, violent crime, the pervasive accumulation (and the sometimes actual use) of excessively (and even catastrophically) destructive weapons, the tendency of national States to avoid cooperation and mutual accommodation, the tendency of national States (or factions within national States) to use war (and otherwise unspeakably dark-minded violence) as a method for achieving the goals of national and *otherwise* culturally idealized policies . . ." The list could be continued; it is long and somber. As we have seen, this scenario of business as usual leads to a dead end.

Other species went toward and into extinction through little or no fault of their own: the environment around them changed, or other species invaded their niche. We do not have more powerful species to contend with, but our environment is changing and may do so irreversibly. The planetary environment is changing because we are changing it. *Homo* the wise, the knower, is outsmarting himself. He is creating untenable conditions in the biosphere and stressful and potentially catastrophic conditions in the sociosphere.

What makes *Homo* create such conditions? Not instincts: those are oriented toward individual and collective survival. *Homo sapiens,* like other species, most notably the higher apes, possess "hard-wired" instincts that make them into adapted social beings. Chimpanzees who cannot swim have sometimes drowned in the moats of zoos trying to save other chimpanzees who have fallen into the water. Rhesus monkeys have been known to starve themselves for days when they could get food only by giving an electric shock to a companion. When male chimps keep fighting, female chimps have been known to take the stones out of their hands, and if they fail to make up after a fight, the females often attempt to bring them together.

Such behaviors go beyond personal benefits, or even interpersonal relationships; they are undertaken for the greater good of the community. As biologists Marc Hauser and Frans de Waal argue, these behaviors are precursors of human morality. They are testimony that higher species possess instincts oriented toward the well-being of their family, group, or tribe and, in the last count, of their species.

We humans have very likely inherited these instincts from our early ancestors, for they were necessary to ensure that our ancestors could survive long enough to take the next step in species' evolution. For better or for worse, their next step was evolution into modern humans. But the instincts of modern humans no longer govern behavior: they have been overlaid by the specifically human ability to engage in reasoning above and beyond the dictates of inborn leanings and preferences. Human reason has the freedom to ignore, and even go counter to, genetically coded patterns of behavior. It is capable of altruism and empathy but also of egoism and intolerance. Today it is the egoic, shortsighted rationality of modern man that guides his steps. It is what creates his values, governs his perceptions, and creates the complex superstructure proudly called modern civilization. This kind of rationality is now testing the limits of the viability of our species.

The unique freedom of *Homo* to err is also his unique salvation. For what has been repressed has not been lost; what is now ignored is not beyond recovery. It is not raw instinct that we need to recover, for it alone is not sufficient to turn around the current rush toward extinction. Deep insight welling from the most basic instincts of our species for collective survival is what we need, for that—combined with the basic wisdom deposited in the great spiritual traditions and rediscovered at the cutting edge of the sciences—can lead us to a condition that is truly viable: to a civilization that is holistic, peaceful, and sustainable.

Deep insight is a reliable resource, for it is the purest contact we can have with reality, contact uncorrupted by pretension and unadorned by sophistry. Were it not for the emergence of such insight at crucial epochs in our history, we would not still be here. But in our history such insight has emerged again and again, and so we are here today. And because it is emerging again, we have a chance of being here tomorrow.

The insight Adi Da expresses is that we are not only threatened, we can also be saved. The threats come from our egoic separateness, and the salvation from the rediscovery of our unity: the unity that is prior to all other facts and considerations. It is there: it is a fact. Unfortunately for modern man, it is a nearly forgotten fact. But, fortunately for all who can move with the times, it is a fact that can be, and is now being, recalled and rediscovered. It is recalled by spiritual masters and rediscovered by front-line thinkers and scientists.

Scientists now know (as we shall see in part 2) that particles are entangled—nonlocally connected—with each other throughout space: they have a prior unity that is active and manifest. Living things of all kinds are nonlocally connected throughout the biosphere; theirs is a subtle connection that is likewise active and real, although we have only recently discovered it. As anthropologists have found, so-called primitive (but in many respects highly sophisticated) tribes are also nonlocally—telepathically—connected with one another, with their homeland, and with their environment. They have not repressed their prior unity. But modern humans have repressed the recognition of our prior unity and then, emboldened by misguided rationality, denied its very existence. We are now witnessing the consequences: nature overexploited and despoiled, thousands of millions pressed into deep and seemingly hopeless poverty, and the human world fragmented into "me" and "others," "my country" and "foreign countries," and "my company" and "competitors."

Return to unity—to seamless wholeness, as in the legendary paradisiacal state Utopia? No. It is the uncompromising requirement of

Homo's physical, biological, and sociopsychological survival. Will this requirement be met? Time will tell, and it will not be long before it tells.

I strongly believe that the answer will be "*Yes.*" We are not alone. Not only are we not alone in the universe—for there is an overwhelming probability that many civilizations exist on some of the innumerable planets of this and billions of other galaxies—we are not alone because there are unseen yet now increasingly manifest forces guiding our destiny. The evidence speaks loud and clear. Voices of true reason rise, a new spirituality evolves, spiritual people tell us that a higher frequency of radiation emerges on the planet. The insight to which Adi Da gives voice is the same insight that is dawning on increasing numbers of people: a decade or two ago thousands, now millions.

The transformation of the human species has begun. A new epidemic is spreading among us: more and more people are infected by the recognition of their unity. The fragmentation of human communities and the separation of man and nature were but an interlude in human history, and that interlude is now coming to a close. We are recovering our unity, not by returning to a prior culture and consciousness, but by moving beyond the fragmented, egoic civilization that has dominated humankind for the past two centuries—moving toward a cooperative world constituted by free people who are capable of representing the interest of the human species.

It is high time to move on: the hour of decision approaches. If a critical mass among us recovers the lived experience and attains the felt realization of our prior unity we shall take action, and we can then await the hour of decision with confidence. The spread of messages coming from the deepest intuitions of which our species is capable is both the means of achieving this paramount condition and an indication that achieving it is not a question of serendipity. It is the fulfillment of our species' deep-seated drive to safeguard and develop the consciousness that is both our blessing and our privilege—and our ineluctable responsibility to use for the benefit of all people and all things that live on Earth.

Part Two

PARADIGM SHIFT
IN
SCIENCE

NINE

The Cosmic Plenum

THE NEW FUNDAMENTAL
CONCEPT OF REALITY

*The quantum shift in the global brain embraces both our experi-
ence and our understanding of the world. Not only is the world
around us in radical transformation, but also our understanding of
the fundamental nature of that world—and of the universe itself—
is shifting. This is important, for our world is an integral part of
the universe; its laws and processes apply on Earth the same as
everywhere else. A sound orientation in our world calls for a sound
understanding of the universe.*

*Science's understanding of the fundamental nature of the uni-
verse is different from what most people believe it is. The universe
does not consist of bits of matter moving about in space and time.
Matter, in the last analysis, is a bound form of energy, and space
and time are an integral dynamic element, interacting with matter
and energy in all its forms. Moreover, the various forms of energy
emerge from and are embedded in a fundamental field or medium
that was not part of the conventional world picture. This "deep
basement" of the universe is variously called quantum vacuum, unified vacuum, physical spacetime, hyperspace, or nuether. Despite
these abstruse names, its existence is not merely theoretical and—
notwithstanding the implication of its names—it is not a vacuum*

and not just space. It is an energy- and information-filled "cosmic plenum," the womb of all that exists, and the background of all that happens, in space and time.

The latest advances in the natural sciences give us a new picture of reality. The universe does not consist of matter in three-dimensional space and time. The classical concept of "mass-points" governed by rigorous universal forces of cause and effect in passive and uniform three-dimensional space and in likewise passive and uniformly flowing time is superseded. In the new concept the universe is an organically interconnected evolving system. Its origins, as well as its fields and forces of interaction, are traced to a cosmically extended fundamental medium.

Theories of this deep level of physical reality, like other theories in science, do not claim eternal validity: they are open to revision. Yet the concept of reality now emerging at the horizons of the natural sciences may not actually be in need of revision in regard to its affirmation of what the universe is *not*. The new physics can affirm with a high degree of confidence that ours is not a universe where matter moves about in neutral space, governed by simple rules of cause and effect. Instead, our best insight is that ours is an evolving, instantly and enduringly interconnected fundamentally integral reality, a universe embedded in a dynamic and physically real medium that subtends the familiar world of three-dimensional space and correlated time. What we call "matter" is a waveform energy pattern in this medium. Let us now take a deeper look at the findings that ground this paradigm shift in the physical sciences.

Quanta—the particles we used to think of as the building blocks of material reality—turned out to be more like waves than like corpuscles. The wave nature of quanta was shown recently by an ingenious experiment by the Iranian-American physicist Shahriar Afshar. The experiment, a modified version of the familiar "dual-slit experiment" pioneered by Thomas Young at the beginning of the nineteenth century, demonstrated that of the dual aspects of particles—wavelike and corpuscular—the wave aspect is fundamental. (The dual-slit experiment will be discussed in more detail in the next chapter.) Even when the experiment is so set up that the corpuscular aspect of a particle is observed, the wave aspect is still there, as shown by the interference pattern formed by the waves that builds up on the screen; it does not disappear when the photons—seemingly discrete entities—pass one or the other of the two slits, presumably one by one. Niels Bohr's celebrated "complementary principle"—which claims that a particle can behave as a corpuscle or a wave but never both at the same time—suggests that the alternative properties of the particle are *complementary*: although not appearing singly, together they fully describe the particle's state. But in Afshar's experiment the wave aspect *is* present even when the corpuscular aspect is observed, whereas the corpuscular aspect is *not* present when the wave aspect is queried.

The implications of such findings are revolutionary. Although what we perceive with our senses is solid matter moving about in empty space, in reality the material universe, including particles, stars, planets, rocks, and living organisms, is not material: matterlike things are standing, propagating, and interacting waves in a subtending medium.

The concept of a physical field that subtends the three-dimensional world of space and time surfaced in the course of the twentieth century. Until the beginning of that century space was believed to be filled with a luminiferous ether that produces friction when bodies move through it. When in the Michelson-Morley experiments such friction failed to materialize, the ether was removed from the physicists' world picture. The absolute vacuum took its place. However, the vacuum turned out to be far from empty space. In the "grand unified theories" (GUTs) developed in the second half of the twentieth century the concept of the vacuum transformed from empty space into the medium that carries the zero-point field, or ZPF (so-called because in this field energies prove to be present even when all classical forms of energy vanish: at the absolute zero of temperature). Ever more interactions have come to light between this fundamental field and observed things and processes. In the 1960s Paul Dirac showed that fluctuations in fermion fields produce a polarization of the ZPF of the vacuum, whereby the vacuum in turn affects the particles' mass, charge, spin, or angular momentum. At around the same time Andrei Sakharov proposed that relativistic phenomena (the slowing down of clocks and the shrinking of yardsticks near the speed of light) are the result of effects induced in the vacuum due to the shielding of the zero-point field by charged particles. In current super grand unified theories (super GUTs) all the forces and fields of the universe are traced to origins in the "unified vacuum."

However, in the technical framework of quantum field theory the vacuum is not a part of physical reality. It is a theoretical artifact, a requirement of the mathematics of the field theory. The insight that the vacuum is a real, and indeed fundamental, medium does not derive from the mathematics of quantum field theory but from significant, if necessarily indirect, evidence accumulated independently in a vast variety of observations.

The evidence for the realistic concept is of two kinds. One comes from the new physics and cosmology: more and more scientists are coming to the conclusion that the level of quanta, and of spacetime itself, is not the ultimate level in the universe. There is also a level below quanta and spacetime, a level from which spacetime and the quanta that populate it have emerged. The other kind of evidence concerns the observation that quanta and the things composed of quanta (organisms and minds included) are intrinsically and, as it appears, "nonlocally" connected. This raises the possibility that the fundamental level of the universe is not merely at the origin of the things that populate space and time but is also the medium that interconnects them. (We shall return to this argument and the evidence for it in chapter 10.)

These strands of evidence suggest that the mathematical vacuum concept of quantum field theory is not a full description of what is still generally (and now misleadingly) called "vacuum." There are two different kinds of vacuum concepts in contemporary physics: the by now "classical" concept of the vacuum as a theoretical construct and the emerging, revolutionary concept of the vacuum as a fundamental medium in the universe. In order to avoid mixing up these different concepts, it is advisable to surrender the label "vacuum" in regard to the new concept; it does not fit it in any case. Thus, in this chapter, and henceforth throughout this book, we shall refer to the vacuum in the emerging context as the "cosmic plenum" and leave "quantum vacuum" to stand for the abstract technical concept of quantum field theory.

Former MIT physicist Milo Wolff summed up the radical implications of the vacuum viewed as a fundamental medium. According to Wolff, this medium is the single source of matter and natural law in the universe. His conclusion: "Since the waves of each particle are intermingled with the waves of other matter and all contribute to the density of the medium, it follows that every charged particle is part of the universe and the universe is part of each charged particle."

A theory that embraces the vacuum as the cosmic plenum completes and complements Einstein's theory of relativity (although it places in doubt one of its pillars, the constancy of the speed of light). Relativity theory views spacetime as relative and dynamic, interacting with matter and energy. It is the "background" against which the events of the manifest world unfold. But the origins of this background are not accounted for in relativity theory: spacetime is simply "given," together with matter and energy. This is much the same in the currently elaborated TOEs (so-called theories of everything). TOEs would be truly theories of *everything* only if they were "background independent"—that is, if they did not merely assume the presence of spacetime but showed how it arose in the universe. The TOEs developed to date, based for the most part on string and superstring theories, cannot do this. Even the highly accomplished M-Brane version of superstring theory advanced by Edward Witten in 1995 fails to provide an answer (this theory has other problems as well: it also does not account for the existence of dark matter and calls for eleven dimensions in the universe rather than the experienced three and relativity theory's four).

The current impasse points toward the need to recognize a deeper floor of the universe. "If we are ever going to find an element in nature that explains space and time," John Wheeler asserts, "we surely have to find something that is deeper than space or time—something that itself has no location in space or time."

There is also independent evidence speaking to the reality of a cosmic plenum. Since Einstein published his general theory of relativity in

1915, important evidence has come to light regarding the existence of a medium that would underlie the observable manifestations of the universe. Initially this cosmic medium was identified with space itself. In the nineteenth century, mathematician William Clifford suggested that small portions of space are analogous to little hills on a surface that is, on average, flat; the ordinary forces of geometry do not hold for them. The property of space to be curved or distorted, he said, is continually being passed on from one portion of space to another after the manner of a wave. This variation in the curvature of space is what really happens when matter moves. Thus in the physical world nothing else takes place but this wavelike variation.

In his 1930 paper "The Concept of Space" Einstein himself noted, "We have now come to the conclusion that space is the primary thing and matter only secondary; we may say that space, in revenge for its former inferior position, is now eating up matter." A few years following the publication of Einstein's thought, Erwin Schrödinger restated the basic insight. "What we observe as material bodies and forces," he noted, "are nothing but shapes and variations in the structure of space."

A further strand of evidence concerns the propagation of light in empty space. There is significant evidence that relativity theory's claim that the speed of light is constant in a vacuum is not universally true. Already in 1913 G. Sagnac provided experimental proof that the speed of light varies with the clockwise and counterclockwise rotation of the light source, but this finding was almost entirely disregarded by the physics community. As of mid-century other investigators, notably Herbert Ives and Ernest Silvertooth, provided experimental evidence for variations in the speed of light in what can no longer be considered empty space. Then, in independent experiments carried out in the late 1990s, Lene Hau and M. Fleischauer showed that light slows down, and ultimately freezes, at temperatures approaching absolute zero. The observed constant speed of 299,792,458 meters per second may be valid only as the general case in this universe.

More and more theories ascribe physical properties to space, more exactly to the field or medium that subtends space. The Italian physicists

Davide Fiscaletti and Amrit Sorli suggested that the stage on which natural phenomena take place is an atemporal four-dimensional physical space (ATPS). "Empty" space, as well as the manifest quanta that make up observable reality, are constituted of "quanta of space" (QS) within the ATPS; the QS are the fundamental building blocks of physical reality. They are of Planck length; they vibrate at a "basic frequency," while the quanta of the manifest world vibrate at lower frequencies. Each manifest quantum is the result of the interaction of energy in the "entropy-state" with one or more QS—the latter are in a nonentropy-state. Quanta devoid of internal structure (such as quarks, leptons, and intermediate bosons) are the result of interaction with one quantum of space; particles endowed with internal structure (baryons constituted of three quarks and mesons made up of a quark-antiquark pair) are the product of interaction with several QS.

Fiscaletti and Sorli maintain that, as both quanta and fields are special states of atemporal physical space, the latter is ontologically primary. The universe, they conclude, is an atemporal phenomenon, and the Planck-length quanta of space are its elementary constituents.

While experimentation and theory-building continues, it is already safe to say that truly empty space is relegated to history. The reality recognized at the frontiers of physics is a cosmic plenum filled with universal forces and virtual particles. The observable and measurable world of particles and particle interactions is a subset of this plenum. At the birth of this universe particles and the entire interacting world of particles emerged out of the cosmic plenum, and it is into this plenum that they die back at the final evaporation of galaxy-size black holes.

The discovery of a level of reality that subtends the manifest world is a logical step in the development of science. The world recognized by scientists is becoming progressively vaster. In the middle of the twentieth century our galaxy was thought to be the entire universe; today we know that there is also a metagalaxy, containing billions of other galaxies. Further, we also know that the metagalaxy is merely *our* universe: there may be millions if not billions of other universes in the meta-universe or "Metaverse."

Not only the horizons but also the floor of the universe is being progressively extended. Underlying the manifest three-dimensional world of particles, forces, and interactions there is a world that does not contain energy and matter in the known form, nor does it include space and time in the accepted sense. This deeper basement is the cosmic plenum. It transmits photons and bosons—the wave propagations that we know as light and force—and constitutes the common substrate of all the universes that evolve and devolve in the Metaverse.

TEN

Nonlocal Coherence

THE NEW CONCEPT OF
MANIFEST REALITY

The "deep floor" of the universe—the cosmic plenum—is not the only surprising discovery at the cutting edge of the sciences. It turns out that the coherence of the universe is also far greater than was hitherto thought. This coherence is quasi-instantaneous throughout space, and it is enduring in time. It is "anomalous" with respect to the understanding we have had of the nature of connections in space and time.

The two discoveries—the cosmic plenum and space- and time-transcending coherence—are not unrelated. Indeed, it is reasonable to suppose that the cosmic plenum is the medium that connects things throughout space and time and creates the manifest kind of coherence among them.

The universe is wholly and enduringly coherent. The phenomenon of coherence is well known: it concerns light waves that have a constant difference in phase. In a condition of coherence phase relations remain constant and processes and rhythms are harmonized. Ordinary light sources are coherent over a few yards; lasers, microwaves, and other technological light sources remain coherent for considerably greater distances. But the kind of coherence that is coming to light in various branches of the empirical sciences is more complex and significant. It indicates a quasi-instant connection among the parts or elements of a thing, whether that thing is a quantum, an atom, an organism, or a galaxy. This kind of coherence surfaces in fields as diverse as quantum physics, biology, cosmology, and brain and consciousness research.

COHERENCE IN THE DOMAIN OF THE QUANTUM

The first experiment to demonstrate the coherence of quanta was conducted by Thomas Young in 1801. In Young's "double-slit experiment" (noted in the previous chapter) light is allowed to pass through a filtering screen with two slits. The beam of light is extremely weak so that each light particle—photon—is emitted separately (in current versions of the experiment, lasers are used for this purpose). The individually emitted photons pass through the first screen to another screen that is placed behind the first, to register the photons that traverse the first. Then, just as when water is allowed to flow through a small hole, the light beam made up of the photons fans out and forms a diffraction pattern. The pattern shows the wave-aspect of light and is not paradoxical in itself. The paradox comes in when a second slit is opened in the first screen. This results in a superposition of two diffraction patterns although each photon was emitted individually and has presumably traveled through only one of the two slits. Yet the waves behind the slits form a characteristic interference pattern, canceling each other when their phase difference is one hundred and eighty degrees and reinforcing each other when they are in phase. Although they pass through different slits, they interact instantly with each other. As waves they could pass through both slits, but as particles that were emitted individually they have the properties of corpuscles and can pass through only one of the slits.

John Wheeler's "split-beam" experiment discloses the same effect. In this experiment, too, photons are emitted one at a time, and they travel from the emitting gun to a detector. A half-silvered mirror is inserted along the photon's path, splitting the beam. On the average, one in every two photons is expected to pass through the mirror and one in every two to be deflected by it. This expectation is borne out: photon counters inserted behind the half-silvered mirror and at right angles to it register an approximately equal number of photons. But when a second half-silvered mirror is inserted in the path of the photons undeflected by the first mirror, all photons arrive at one destination and none at the other. This suggests that the kind of interference that was noted in the double-slit experiment also occurs in the split-beam experiment. Above one of the mirrors the interference is destructive (the phase difference between the photons is 180 degrees), so that the photons, as waves, cancel each other. Below the other mirror the interference is constructive (since the wave phase of the photons is the same) and the photon waves reinforce each other.

The interference patterns of photons emitted moments apart in the laboratory are also observed when the photons are emitted at considerable distances from the observer and at considerable intervals of time. In the "cosmological" version of the split-beam experiment the observed photons are those that were emitted by a distant star; in one case, by the double quasar known as 0957+516A,B. This distant "quasi-stellar object" appears to be two objects but is in fact one, its double image being due to the deflection of some of its light by an intervening galaxy. The photons of the light beam deflected by the intervening galaxy have been on the way fifty thousand years longer than the photons in the undeflected beam. Yet the photons, originating billions of years ago and arriving with an interval of fifty thousand years between them, interfere with each other similarly to those emitted seconds apart in the same laboratory.

The coherence of quanta is further shown by experiments with so-called which-path detectors (detectors that "label" the individually emitted photons in order to identify which path they have taken, which slit they have passed through). When the which-path detectors are active, the quanta begin to behave as classical objects: their interference is damped (physicists note that the "interference fringes" that build up on the screen diminish). In the experiment conducted by Eyal Buks, Mordehai Heiblum, and collaborators at Israel's Weizmann Institute, a device less than one micrometer in size created a stream of electrons across a barrier on one of the two paths. The paths focused the electron streams and made possible the measuring of the level of interference of the electrons in the streams. The investigators found that the higher the detector is tuned for sensitivity, the less the interference is pronounced. With the detector turned on for both paths, the interference fringes disappear entirely.

Other experiments show that the interference fringes disappear as soon as the detector is installed, even if it is not turned on. In Leonard Mandel's optical-interference experiment of 1991 two beams of laser light were generated and allowed to interfere. When a detector was present that enabled the path of the light to be determined, the interference fringes disappeared. But the fringes disappeared regardless of whether or not the determination was carried out. This showed that the very possibility of "which-path detection" destroys the interference pattern.

The above finding was confirmed in 1998 by Dürr, Nunn, and Rempe in an experiment where interference fringes were produced by the diffraction of a beam of cold atoms by standing waves of light. When no attempt was made to detect which path the atoms were taking, the interferometer displayed fringes of high contrast. However, when information was encoded within the atoms as to the path they took, the fringes vanished. The labeling of the paths did not need to be read out to produce the disappearance of the interference pattern; it was enough that the atoms were labeled so that this information could be read out.

It appears that not only do individually emitted, and hence presumably corpuscular, particles or atoms interfere with each other as waves, but a which-path detecting apparatus is also coherently coupled with the stream of particles or atoms to which the apparatus is tuned. These findings bear out the concept of "entanglement" suggested by Schrödinger in 1935. Quanta occupy collective quantum states. The quantum states of all particles within a system of coordinates are "superposed" so that it is not the property of a single particle that carries information but the state of the system of coordinates in which the particle is embedded. In that system the individual particles are intrinsically "entangled" with each other. The superposed wave function of the whole system describes the state of each particle in it.

COHERENCE IN THE COSMOS

The coherence of cosmic ratios. A number of noteworthy coincidences have come to light regarding the physical parameters of the universe. In the 1930s Sir Arthur Eddington and Paul Dirac noted that the ratio of the electric force to the gravitational force is approximately 10^{40}, and the ratio of the observable size of the universe to the size of elementary particles is likewise around 10^{40}. This is all the more surprising, given that the former ratio should be unchanging (the two forces are assumed to be constant), whereas the latter is changing (since the universe is expanding). In his "large number hypothesis," Dirac speculated that the agreement of these ratios, the one variable, the other not, is not merely a temporary coincidence. But if the coincidence is more than temporary, either the universe is not expanding or the force of gravitation varies in accordance with its expansion.

Additional coincidences involve the ratio of elementary particles to the Planck length (which is 10^{20}) and the number of nucleons in the universe ("Eddington's number," approximately 2×10^{79}). These are very large numbers, yet harmonic numbers can be constructed from them. (Eddington's number, for example, is roughly equal to the square of 10^{40}.)

Recently physicist Lee Smolins discovered additional numerical coincidences. Observations indicate that the cosmic microwave background radiation is dominated by a large peak followed by smaller harmonic peaks. The series ends at the longest wavelength, which Smolins terms R. When R is divided by the speed of light, we obtain a measure of time that agrees with the age of the universe. When in turn the speed of light is divided by R, we get a frequency that equates to one cycle over the age of the universe. And the speed of light squared and divided by R (c^2/R) gives a measure of acceleration in the universe that corresponds to the acceleration observed and attributed to dark energy!

Menas Kafatos and Robert Nadeau showed that many of the coincidences can on the one hand be interpreted in terms of the relationship between the masses of elementary particles and the total number of nucleons in the universe, and on the other in terms of the relationship between the gravitational constant, the charge of the electron, Planck's constant, and the speed of light. Scale-invariant relationships appear. The physical parameters of the universe turn out to be generally proportional to its overall dimensions.

The coherence of the universal constants. Coherence among the numerical parameters of the universe is complemented by coherence among the values of the universal laws that govern interaction in space and time: the "universal constants." The coherence of the constants involves upward of thirty factors and considerable accuracy. For example, if the expansion rate of the early universe had been one-billionth less than it was, the universe would have recollapsed almost immediately; if it had been one-billionth more, it would have flown apart so fast that it could produce only dilute, cold gases. A similarly small difference in the strength of the electromagnetic field relative to the gravitational field would have prevented the existence of hot and stable stars like the Sun, and hence the evolution of life on planets that are physically capable of supporting life. Moreover, if the difference between the mass of the neutron and the proton was not precisely twice the mass of the electron, no substantial chemical reactions could take place, and if the electric charge of electrons and protons did not balance precisely, all configurations of matter would be unstable and the universe would consist merely of radiation and a relatively uniform mixture of gases.

That the large-scale coherence of the universe would be merely a vast series of "coincidences" is extremely improbable. It appears that already at the universe's birth, the Big Bang that created the particle/antiparticle pairs—whose excess particles furnish the substance of the universe—was precisely tuned to produce constants that permitted the subsequent evolution of systems of growing complexity.

COHERENCE IN THE BIOSPHERE

Quantum-type coherence in the organism. Quanta appear to be intrinsically coherent, but larger-scale systems were assumed to exist in so-called classical states—states of "de-coherence." However, this is not the case. Complex molecules, cells, and living organisms turned out to exhibit quantum-type processes on the macroscopic scale. This was demonstrated in 1995 by Eric A. Cornell, Wolfgang Ketterle, and Carl E. Wieman, in experiments for which they received the 2001 Nobel Prize in physics. The experiments show that, under certain conditions, seemingly separate particles and atoms interpenetrate as waves. For example, rubidium and sodium atoms behave not as classical particles but as nonlocal quantum waves, penetrating throughout the given system and forming interference patterns.

In 1999 atoms of an extremely heavy isotope of carbon known as "buckminsterfullerene" were shown to be capable of entanglement: they proved to have wave properties as well as corpuscular properties. By 2005 even complex organic molecules could be entangled, and some have been "teleported" over considerable distances in the manner of subatomic particles. In 2007 biophysicists Gregory Engel and collaborators reported experiments that show that quantum-type coherence is already present in green sulphur bacteria: it acts as an energy "wire" that connects the light-harvesting chromosome to the bacterial reaction center. Without the wavelike energy transfer created through quantum coherence, the efficient kind of photosynthesis that had allowed life to get started on this planet could not have taken place; there would not be life on Earth.

Complex organisms could not have evolved and could not function in the absence of the nonlocal forms of coherence. The human body, for example, consists of 10^{14} (100,000,000,000,000) cells, and each cell produces 10,000 bioelectrochemical reactions every second. These all need to be quasi-instantly and dependably correlated. Moreover every night 10^{12} (1,000,000,000,000) cells die and are replaced by roughly the same number. The coordination of this vast number of cells in the organism and their complex electromagnetic and chemical signaling cannot be explained by physical and chemical interactions alone. Although some signaling—for example, by control genes—is remarkably efficient, the speed with which activating processes spread in the body, as well as the complexity of these processes, makes explanation in reference to biophysics and chemistry alone insufficient. The conduction of signals through the nervous system, for example, cannot proceed faster than about sixty-six feet per second, and it cannot carry a large number of diverse signals at the same time. Yet there are quasi-instant, nonlinear, heterogeneous, and multidimensional correlations among all cells in the organism, conveyed through organs and entire organ systems.

Correlations of this kind suggest the form of coherence observed in the domain of the quantum. If distant cells, molecules, and molecular assemblies are to resonate at the same or compatible frequencies, they must resonate in phase: the same wave function must apply to them. This also holds true in regard to the coupling of frequencies among the assemblies: if faster and slower reactions are to accommodate themselves within a coherent overall process, the respective wave functions must coincide.

The latest findings show that the living organism is a coherent system—more exactly, a macroscopic quantum system. In the language of physics, it is governed by an integral "macroscopic wave function."

The coherent evolution of organisms. The fact that biological organisms could evolve on this planet is a strong if indirect indication of an embracing level of coherence in the living world. This coherence embraces the genome and the phenome within organisms and organisms and their environments in the biosphere.

There is statistical as well as experimental evidence that the genetic information encoded in the organism and the phenome (the biological organism) that results from this information are interconnected. Contrary to the classical Darwinian doctrine, the genome does not mutate purely randomly, unaffected by the changing states of the phenome. This is important, for in the absence of such connection the evolution of complex organisms would be astronomically improbable. The "search space" of genetic rearrangements is so enormous that random processes would take incomparably longer to produce viable new species than the time that was available for evolution on Earth. It is not enough for genetic rearrangements to produce one or a few positive changes in a species;

they must produce the full set. The evolution of feathers, for example, does not produce a reptile that can fly: radical changes in musculature and bone structure are also required, along with a faster metabolism to power sustained flight. The development of the eye requires thousands of mutations finely coordinated with one another. Yet the probability of a single mutation producing positive results is negligible: statistically only one mutation in 20 million is likely to be viable. By itself, each mutation is likely to make the phenome less rather than more fit, and if so, it will be eliminated in time by natural selection.

An additional factor speaking against the thesis of random mutations producing viable organisms is the possibility that complex organisms are "irreducibly complex." The parts of an irreducibly complex organism are interrelated in such a way that removing any one part destroys the function of the whole. Thus to mutate an irreducibly complex system into a viable system every part has to be kept in a functional relationship with every other part throughout the process. According to Michael Behe, this level of precision is unlikely to be achieved by random piecemeal modifications in the genetic pool of complex biological organisms.

COHERENCE IN THE DOMAIN OF BRAIN AND MIND

At the cutting edge of brain and consciousness research a significant body of evidence has surfaced showing that the brain functions of different individuals can achieve coherence even when the individuals are not in an ordinary form of contact with each other. The space- and time-transcending coherence among the consciousnesses of individual humans is first of all the nonlocal coherence of their brains and bodies.

Telepathic, remote-viewing, and telesomatic phenomena have been subjected to increasingly rigorous experiments. The evidence regarding the phenomenon of "twin pain" (where one of a pair of identical twins intuits or feels the pain or trauma of the other) has been exhaustively investigated; it appears to occur in about 25 percent of identical twins. Spontaneous coherence in brain functions occurs also among genetically unrelated individuals. Laboratory tests show that personal contact, or an emotional bond, are often sufficient to produce the transfer of stimuli among pairs of subjects.

At the National University of Mexico Jacobo Grinberg-Zylberbaum performed more than fifty controlled stimuli-transfer experiments. He paired his subjects inside soundproof Faraday cages and asked them to meditate together for twenty minutes. Then he placed them in separate Faraday cages where one subject was stimulated and the other not. In double-blind experiments the stimulated subject received stimuli (such as flashes of light, sounds, or short, intense, but not painful, electric shocks to the index and ring fingers of the right hand) at random intervals. The electroencephalograph (EEG) brain-wave records of both subjects were then synchronized and examined for "normal" potentials evoked in the stimulated subject and "transferred" potentials in the nonstimulated subject. Transferred potentials appeared consistently in about 25 percent of the cases. In a limited way, Grinberg-Zylberbaum could also replicate these results: when one individual exhibited the transferred potentials in one experiment, he or she usually exhibited them in subsequent experiments as well.

A related ability of individuals is to achieve a high level of spontaneous synchronization of their cerebral functions. A series of experiments carried out by Italian physician and brain researcher Nitamo Montecucco shows that in deep meditation the left and right hemispheres of the brain manifest identical wave patterns. More remarkably, the left and right hemispheres of *different* subjects become synchronized. In one test, eleven out of twelve meditators achieved 98 percent synchronization of the full spectrum of their EEG waves in the complete absence of sensory input. The experiment was repeated in the framework of the Global Peace Meditation/Prayer Day on the 20th of May 2007, when over a million people joined together in sixty-five countries to meditate or pray for peace (see "The Objectives of the Global Peace Meditation/ Prayer Days" in part 3). The events were carefully synchronized and focused on three time zones.

The brain-synchronization experiment was conducted in the time zone where it coincided with meditations in Europe. It involved eight meditators in the town of Bagni di Lucca in Tuscany and eight in the city of Milan. The EEG waves of the subjects were monitored by electrodes on their heads; a computer calculated the level of synchronization among the waves. It turned out that *within* each of the groups the level of synchronization was not as high as in previous experiments (this was very likely because of interruptions due to problems with the equipment), but the correlation *between* the two groups, hundreds of miles apart and not in any ordinary form of contact with each other, was significant: it reached the value of 0.53 on the scale of probability, entirely beyond the pale of mere serendipity.

Additional evidence of the transmission of physical effect between individuals in the absence of sensory contact is furnished by spiritual healing. Psychiatrist Daniel Benor analyzed hundreds of cases of controlled experiments in spiritual and nonlocal healing and found significant evidence of positive therapeutic effect.

The transfer of effect from healer to patient can be monitored and measured: it shows up in their EEG waves. C. Maxwell Cade of the Institute of Electrical Engineers in England tested the EEG patterns of over 3,000 people in various states of consciousness. He found five characteristic states, where each state manifests a specific combination of wave frequencies (the known frequencies are beta, with a range between 13 and 30 Hz; alpha, ranging from 8 to 13 Hz; theta, between 4 and 7 Hz; and delta, in the range of 0.5 to 4 Hz). The normal waking state is almost entirely in the range of beta. Alpha occurs in meditation and restful states, theta in half-awake or dreaming-sleep states, and delta in profound dreamless sleep. Healers function typically in what Cade called the "fifth" state, consisting of a moderate amount of beta and theta, wide alpha, and no delta (though the latter finding has exceptions, as we shall see). Cade found that in the process of healing the healer induces his or her characteristic fifth state pattern in the patient.

The transfer of brain-wave pattern can also occur when healer and patient are in separate locations and neither hear nor see each other. This was shown in an experiment witnessed by this writer in southern

Germany in the spring of 2001. At a seminar attended by about a hundred people, Dr. Günter Haffelder, head of the Institute for Communication and Brain Research of Stuttgart, measured the EEG patterns of Dr. Mária Sági (the scientific secretary of the Club of Budapest who is also a gifted natural healer) together with those of a young man who volunteered from among the participants. The latter remained in the seminar hall while Dr. Sági was taken to a separate room. Both healer and healee were wired with electrodes, and their EEG patterns were displayed on a monitor in the hall. During the time Dr. Sági was diagnosing and treating the young man, her EEG waves were in the theta range and dipped even into the deeper delta region with a few eruptions of wave amplitude. The EEG of the young man, who sat in the hall in a light meditative state, exhibited the same pattern with a delay of about two seconds. Yet they had no sensory contact with each other.

ELEVEN

The Akashic Field

THE NEWLY REDISCOVERED CONCEPT OF REALITY

The coherence of parts of a system that are neither contiguous nor connected by known forms of energy and information suggests that there are forms of connection in nature that are not known to mainstream Western science. This is now changing. The recognition of a cosmic plenum—the "deep-floor of the universe"—offers a logical starting point for investigating the physical basis of interconnection wherever it occurs: in the cosmos, in the human world, and even in the domain of mind and consciousness.

FIELDS

The quasi-universal phenomenon of nonlocal coherence suggests that, in addition to harboring the universal and quantum fields, the cosmic plenum also acts as a universal interconnecting field. The concept of a *field* is an important contribution to the conceptual arsenal of science. Objects that exhibit some form of interconnection beyond the immediate effect of physical causality are said to be connected through an underlying field.

Fields are not observable in themselves; they can only be inferred from phenomena taken to be their effect. For example, the gravitational field (G-field) itself cannot be observed: when an object drops to the ground, we observe the object falling but not the field that makes it fall—we see the effect of the G-field but not the G-field. The same applies to the electromagnetic field, where the effect is the transmission of electric and magnetic force, and to the weak and strong nuclear fields where the effect is attraction and repulsion among particles at extreme proximity to each other. The coherence observed in nature indicates that there is yet another field active in the universe: a field that produces quasi-instant connection among separate and even distant objects.

Postulating a universal field to account for observations that would otherwise remain unexplained has well-established precedents in the history of science. In the nineteenth century Faraday discovered that electric and magnetic phenomena imply interaction among individual objects separated by finite distances: he postulated the electromagnetic field (EM-field). Faraday's field was a local field, associated with the given objects; it was Maxwell's insight that the EM-field is universal. This was a revolutionary postulate, for if the electromagnetic field is universal, space can no longer be considered empty and passive: it must be a continuous field conveying electric and magnetic effects.

The field accounting for gravitational attraction among massive particles has a similar history. In Newton's theory gravitation is a local phenomenon, an intrinsic property of objects with mass. Newton, and subsequently Ernst Mach, was deeply puzzled by this property. It was Einstein who resolved the mystery by removing the gravitational force from individual objects: he ascribed it to four-dimensional spacetime. In the general theory of relativity, the G-field is a universal field.

In recent years still another universal field entered the world picture of physics: the Higgs field. Similar to gravitation, the Higgs field has to do with mass, but not with the property of massive objects or with the action of fields on massive particles. The Standard Model of particle physics suggests that the Higgs field is the universal field that *creates* mass in particles.

What these examples tell us is that when phenomena occur that require a physical explanation, the first attempt is to give an explanation specifically related to the entities that manifest the phenomena. However, as theories grow and mature, the explanatory concepts become more general. Electric and magnetic phenomena are now ascribed to the universal EM-field, the mutual attraction of noncontiguous objects is ascribed to the universal G-field, and the presence of mass is ascribed to the universal Higgs field. By the same reasoning it is logical to ascribe the nonlocal coherence observed in nature to an interconnecting field.

THE AKASHIC FIELD

We now take another step toward completing the theory of a nonlocally interconnecting field: we identify it as the Akashic field.

In Indian philosophical traditions, *akasha* is the Sanskrit term for the most fundamental of the five elements of the cosmos, the others being *vata* (air), *agni* (fire), *ap* (water), and *prithivi* (earth). Akasha embraces the properties of all five elements: it is the womb from which everything our senses can perceive has emerged and into which everything will ultimately redescend. Akasha *underlies* all the manifest phenomena of the cosmos; it *becomes* all the manifest phenomena. It is said to be so subtle that it cannot be perceived until it becomes manifest in the world.

Even if our sense organs do not register Akasha, we can reach it through spiritual practice. The ancient Rishis (seers) reached it through a disciplined, spiritual way of life and through yoga. They described their experience and made Akasha an essential element of the philosophy and mythology of India.

In the twentieth century Akasha was brilliantly described by Swami Vivekananda. According to Vivekananda, Akasha is omnipresent and all-penetrating in the universe. Everything evolved out of Akasha. It is Akasha that became the human body, the animal body, the plants, every form that we see, everything that exists. At the beginning of the universe there was only Akasha, and at its end there will be only Akasha again: the solids, the liquids, and the gases all melt back into it. At the next cycle of the universe the energies now active in the universe will start up again, and all things will evolve out of Akasha.

It is now evident why the universal interconnecting field merits the name "Akashic." In the view put forward by Vivekananda, Akasha is functionally equivalent to—that is, for all intents and purposes it is essentially the same as—the cosmic plenum. It is the womb of all "matter" and all "force" in the universe.

A hundred years ago Nikola Tesla spoke of an "original medium" that fills space and compared it to Akasha, the light-carrying ether. In his unpublished 1907 paper "Man's greatest achievement" Tesla wrote that this original medium, a kind of force field, becomes matter when Prana, cosmic energy, acts on it, and when the action ceases, matter vanishes and returns to Akasha. Since this medium fills all of space, everything that takes place in space can be referred to it. Four-dimensional curvature, said Tesla, put forward at the time by Einstein, is not the decisive feature of space.

However, by the end of the first decade of the twentieth century physicists adopted Einstein's mathematically elaborated four-dimensional curved spacetime and, with the exception of a few maverick theoreticians, refused to accept any concept of a space-filling ether, medium, or force field. Tesla's insight fell into disrepute, and then into oblivion. Today it is reconsidered. David Bohm, Harold Puthoff, and a small but growing group of scientists are rediscovering the role of a field that would create coherence throughout the universe.

The Hindu seers were on the right track. There is a deeper reality in the cosmos, a reality that is an Akashic field that connects and creates coherence. This field deserves to join science's G-field (the gravitational field), the EM-field (the electromagnetic field), the Higgs field, and the nuclear and quantum fields as a fundamental feature of the known universe.*

---

*Readers interested in delving deeper into the foundations of the Akashic field can consult the author's *Science and the Akashic Field* (Rochester, Vt.: Inner Traditions 2007).

TWELVE

Metaphysical, Theological, and Ethical Implications

The twin discoveries of a deeper level of reality and the nonlocal coherence of nearly all things that populate the manifest level suggest a radically new vision of the world. This newly rediscovered vision has profound meaning for our understanding of the fundamental nature of the cosmos, as well as of life and mind in the cosmos.

The rediscovered vision suggests, first of all, a different understanding of the first principles that underlie the manifest entities and processes of the universe: a new metaphysics. It also suggests a new concept of the relation of the divine to humans and to the world: a new theology. And it likewise suggests criteria for distinguishing right action from misguided and uninformed action: a new morality.

THE NEW METAPHYSICS

According to Aristotle, metaphysics follows physics as the study of the first principles entailed by our understanding of the nature of reality. This task needs to be reattempted every time there is a fundamental shift in our conception of reality. The following is an outline of a meta-

112

physics that elaborates Alfred North Whitehead's influential process metaphysics in light of the new-paradigm science discussed in the previous chapters.*

Two Domains of Reality

At various levels of evolution the real ("matter-like" but not "material") entities of the universe correspond to what Whitehead called "actual entities" and "societies of actual entities" (or generally, "organisms"). In the new conception the two domains of reality—the domain of actual entities (the "spacetime domain") and the domain of the cosmic plenum (the "field domain")—together constitute reality. They are categorically distinct but not ontologically disparate: they are diachronically as well as synchronically related (that is, they are related sequentially in time as well as simultaneously).

At the logically extrapolated but empirically unverifiable beginning of the cosmic process, only the field domain existed, in a primordial state. This was a spatially and temporally unbounded sea of fluctuating virtual energies: the prespace of the actualized universe. When a region of the cosmic prespace suffered a critical instability, a fraction of the energies thereby liberated became established as actual entities: seemingly independent waves that are nonetheless part of the cosmic plenum in which they appear. The relations of these entities to each other and to the underlying plenum launched the evolutionary process in space and time: it created the first universe in the Metaverse.

Thus, diachronically, the field domain is prior: it is the generative ground of the particles and societies of particles that populate the spacetime domain. Synchronically, the generated entities are linked with their generative ground through an ongoing bidirectional interaction. In one direction the actual entities structure the cosmic plenum in which they

---

*This outline of a post-Whiteheadian metaphysics, like that of the new theology, is concise and inevitably somewhat technical. It is intended primarily for readers with an interest in the first principles that underlie the diverse levels of human experience. Readers with more practical interests can skip to the section dealing with the new morality without loss of meaning or continuity.

are interacting waves. In the other direction the structured plenum "in-forms" the actual entities (waveforms). The interaction leads to the progressive evolution of the spacetime domain and the progressive structuring of the field domain.

The cosmic plenum first generated the particles that are the initial and basic constituents of the spacetime domain and then, in a progressive but intermittent and nonlinear evolutionary process, created sequentially more complex "societies" of actual entities. In the case of our universe, during the 13.7 (or, as a recent finding indicates, 15.8) billion years that have elapsed since its birth in the Big Bang, the two domains have evolved in reciprocal interaction.

The fundamental reality of the cosmos is the field domain. In this domain two kinds of waves are generated: the seemingly independent yet interconnected and interacting waves that make up the actual entities of the spacetime domain and waves that carry information without carrying energy (known in physics as nonvectorial, so-called scalar waves). The latter register and conserve the traces of the actual entities in the field domain.

The field-based waves that create the actual entities of the spacetime domain produce the "hardware" of the universe: its manifest, quasi-material furnishings. The waves that carry information rather than energy are the universe's "software." They govern the behavior and the evolution of the universe's manifest entities. The difference between this universe and any other universe that could or did arise in the Metaverse is the nature of the information carried by nonvectorial waves in the field domain.

Causality

The in-formation of actual entities by the information present in the field interconnects actual entities throughout spacetime. The causal relations that hold sway in that domain are twofold: they include downward as well as upward causation. Classical causation is "upward causation," that is, the causal process whereby a set of parts jointly determine the structure and function of the system formed by them. Evidently, this kind of causation is operative. However, given that all things in spacetime are connected with all other things, there is also downward causation: the causal influence of whole systems on each of their parts.

At the quantum level downward causation creates "entanglement" among particles within their system of coordinates. In the living world downward causation produces coherence and correlation within and among organisms and ecologies. And, at the astronomical level, this form of causation produces coherent evolution among the macrostructures of the universe.

Evolution through Energy and Information

The universe self-evolves through the propagation, transformation, and conservation of *energy* and *information*. In the spacetime domain energy is conserved, but it is degraded when it performs work: free energy becomes progressively unavailable. In the field domain information is not only conserved but is also created through interaction with the spacetime domain. Information accumulates and progressively "in-forms" the evolutionary process.

The in-formation of the spacetime domain by the field domain results in the increase in the complexity of the entities that populate the spacetime domain and the structuring of the field domain in which the entities emerge and through which they are interconnected. The spacetime domain becomes progressively more ordered and at the same time more entropic.*

The persistent in-formation of the spacetime domain through the field domain constitutes an influx of the past into the present. The past is always present. The present is the specious point in the ineluctable flow of time where the cumulative impact of the past sets the context in which the evolutionary path to the future is selected.

---

*The increase in the quantity of information corresponding to the growth of order in a system follows from the widely accepted equivalence of the term for negative entropy in the thermodynamics of Rudolf Clausius and Ludwig Boltzman with the term for information in the information theory of Claude Shannon: $\Delta I = k \log n\ N_1/N_0 = -\Delta S$.

THE NEW THEOLOGY

A consistent and minimally speculative theology can also be developed in reference to the new concept of reality. In this theology God is not separate from creation: God is part of the universe. God's creation is not the universe we observe and inhabit itself but the *potentials* of the universe for its *self-creation.*

The first thing to note is that a primordial creative act is logically required to account for the physical parameters of the universe. The forces and constants of this universe are precisely such that complex systems, including those we call living, can evolve in space and time. That these universal parameters would be the result of a serendipitous selection from among the range of alternatives is contrary to all reasonable estimates of probability. The range of alternatives—that is, the number of physically possible universes—is staggering: according to mathematical physicist Roger Penrose, it is of the order of $10^{10^{123}}$.

If not arising by random selection, the forces and constants that define the evolutionary trajectory of this universe must have been fine-tuned by constraints that were present in the cosmic plenum at the time it arose. This is a scientifically acceptable hypothesis: cosmologists recognize that the Big Bang was not necessarily the beginning of *the* universe, it may have been the beginning only of *our* universe. If our universe was born in the framework of a vaster Metaverse, prior conditions—antecedent universes—could have in-formed its birth.

However, a transfer of information from antecedent universes to our universe does not preempt the need to explain why our universe is so improbably fine-tuned to the evolution of complexity; it only shifts the explanation to the antecedent universe and ultimately to the hypothetical first universe in the Metaverse. If the propensity of our universe to give rise to complex systems was not due to a creative act at its own birth, it must have been due to such an act at the birth of the first universe. The latter assumption is empirically unverifiable but logically required; in its absence we can only give a transcendental explanation of the capacity of our universe to bring forth life and complexity.

Beyond a creative act at the presumed beginning of the universe-generating process, a theology based on the new concept of reality does not call for intervention by a transcendental agency. Known as deism, this doctrine contrasts with the belief system of most religions, which affirm the actuality, or at least the possibility, of ongoing divine intervention. Theologians who search for a reconciliation of the doctrine of ongoing divine intervention with science (Nancey Murphy, Arthur Peacock, and John Polkinghorne, among others) postulate a "top-down" as well as a "bottom-up" form of intervention: the former through the divine provision of an ongoing stream of information that shapes the course of events at both microscopic and macroscopic levels and the latter by a divine shifting of the probabilities of (otherwise random) quantum events on the assumption that variations at that level do not cancel out but produce amplified effects on macroscopic levels. These theologians claim that, through ongoing top-down and bottom-up interventions, God influences the course of evolution without interfering with the laws of nature.

Ongoing divine intervention is not logically required however: the evolution of our universe, and whatever other universes may exist and have existed in the Metaverse, can be understood in reference to natural causes; thus deism is the least speculative hypothesis. In its logically indicated form it maintains that the information that guides the evolutionary process is generated in and by the universe itself. If a creative act endowed the primordial cosmic plenum with the information that governs the interaction of the entities that emerge in the successive universes, the two domains of the cosmos, the energy-processing spacetime domain of actual entities and the information-conserving field domain, interact and give rise to the processes and the products of evolution.

In this deistic theology it is not the "hardware" of the universe that is transcendentally created but the "software": the information that governs the evolutionary process. Given information that creates suitable potentials for correlations among the universe's actual entities, the universe self-evolves. The original presence of this information must be ascribed to a transcendental creative act whereas its effects, growth, and elaboration can be considered to be immanent.

THE NEW MORALITY

The new concept of reality holds important implications for human life and conduct. It provides a scientific framework for the planetary ethic discussed in chapter 6.

A meaningful ethic must be able to distinguish moral action from immoral or merely uninformed action. Essential for distinguishing moral action is the possibility of *choice*. An entity that does not possess the freedom to choose its course of action cannot be said to act morally or immorally: it merely *acts*. This is not the case for human beings who have a significant degree of freedom to choose. Human freedom is significant but not unconditional; we can reconstruct it in reference to the metaphysical implications of the new concept of reality.

Human freedom derives from the energetic interaction of the actual entities of the universe with each other in the spacetime domain and with their in-formational interaction with the field domain. Actual entities have some, even if not total, freedom in regard to deciding the outcome of these interactions. Their level of freedom correlates with their level of evolution. A higher level of freedom is due to the accrued sensitivity of complex entities to the energy and information flows reaching them. This sensitivity is limited in relatively simple entities to basic and relatively unvarying forms of response to stimuli. Complex entities, on the other hand, can be aware of some elements in the incoming energy and information flows and can select their response to them. In humans a high level of awareness constitutes a sophisticated means of perception and cognition, capable of selecting some elements of the flows and ignoring or repressing others.

Awareness of the flows constitutes one aspect of the freedom of complex entities: the aspect of internally guided selectivity. Another aspect of freedom resides in the coupling of reception and response. Unlike in comparatively simple entities, such as atoms, molecules, and the more primitive forms of life, in evolved organisms there is a significant range of variables that intervene between stimulus and response. These "intervening variables" endow evolved organisms with the capacity to choose their response to the energy and information flows that reach them in their milieu. In human beings there is a wide range of variables between stimulus and response, such as tacit preferences, cultural predisposi-tions, and a range of consciously or subconsciously held values and beliefs. The possible responses range from passive inaction to a variety of courses of action. Humans enjoy a twofold freedom: self-determined selectivity in regard to energy flows in the spacetime domain and to "in-formation" through the field domain and the capacity to choose their responses to both.

The freedom enjoyed by humans endows their actions with a moral dimension. Comparatively simple entities such as particles do not sig-nificantly internalize the determinants of a response, consequently they do not have a high degree of freedom. Higher on the scale of biological evolution more and more of the determinants of response are internal-ized, thereby increasing the level of freedom of the species. In such spe-cies the response can be assessed in reference to criteria of morality.

Criteria for Moral Action

Self-selected behavior can be considered moral to the extent that it con-tributes to the coherence of the subject, and to the coherence of the world around the subject. We distinguish internal as well as external forms of coherence.

The internal coherence of a complex entity is the result of the inte-gration of its parts within its overall structure. In a biological organism the level of internal coherence decides the level of organic health. When the organism is internally coherent, its bodily functions are coordinated and effective: it is healthy. When the overall level of coherence is low, the organism's immune system is weak and the organism is subject to mal-function and disease. When an organ is insufficiently coherent with the whole organism it is diseased; when a group of cells is incoherent with the rest of the organism it reproduces without regard to the integrity of the system: it is cancerous.

The external coherence of the entity regards the way it relates to its natural and social environment. In a human being external coherence concerns his or her relations to family, community, enterprise, country, as well as nature.

Internal and external coherence are interrelated and mutually reinforcing. A healthy mind promotes health in the body, and a healthy body sustains mental health. There is evidence that people who are psychologically adjusted within their family, community, and natural environment are less prone to disease, and those who enjoy good health are less likely to engage in antisocial and antiecological behavior.

Creating, or contributing to, coherence in and around us is not an abstract ideal but a norm for optimal functioning. Nature is coherent, and so is the universe. As far as we know there is nothing in the biosphere or in the universe at large that would continuously and purposively hinder coherence. Creating incoherence through unnecessary aggression, violence, and irrational fragmentation and polarization is a uniquely human trait. It is not part of the nature of other entities: nonhuman species are generally coherent, and the behavior of their members is instinctively oriented toward maintaining or improving community and species' coherence. As we have seen, higher apes clearly manifest instinctive behavior that promotes the integrity and well-being, and hence the coherence, of their groups, tribes, and communities.

Human beings can be consciously and willfully destructive of coherence in their bodies as well as in their communities and environments: they have the freedom to act immorally. But humans can also recognize that incoherence-inducing behavior is counterfunctional and aberrant: they can also choose to act morally. A sound planetary ethic rejects incoherence-inducing action in favor of action oriented toward the coherence—the health and integrity—of body and mind, as well as of community and environment.

Moral Choice in Reference to the Minimum and Maximum Codes of a Planetary Ethic

Moral behavior in the above formulation satisfies, and is in fact equivalent to, the minimum and the maximum codes of the planetary ethic put forward in chapter 6.

As we have said, the minimum code requires that one's action should not conflict with the right of others to live and grow: *Live so that others can also live*. Behavior that is moral in the sense that it does not reduce, or otherwise interfere with, the external and internal coherence of other people satisfies the minimum code. The maximum code, in turn, states, *Act so as to further the evolution of a humanly favorable dynamic equilibrium in the biosphere*. Moral action of the kind that enhances the coherence of the human family in its planetary setting satisfies the maximum code. It promotes the persistence of the human species in the framework of the biosphere.

Notwithstanding the different formulations, there is no inconsistency between the minimum and the maximum codes of a planetary ethic and an ethic based on the new concept of reality. The minimum and the maximum codes are the commonsense, practice-oriented formulation of moral behavior. The injunction to choose modes of action that at the least do not reduce or interfere with coherence around us and at best contribute to such coherence is the science-based formulation. In the final count they are equivalent—they express the same concept of morality.

THIRTEEN

The Next Evolution of
Human Consciousness

*Human consciousness is not a permanent fixture: cultural anthro-
pology testifies that it developed gradually in the course of millen-
nia. In the thirty- or fifty-thousand-year history of modern human
beings, the human body did not change significantly, but human
consciousness did. How will it change next? The answer to this
question is of more than theoretical interest: it could decide the
survival of our species.*

A number of thinkers have attempted to define the next step in the evolution of human consciousness. The Indian sage Sri Aurobindo considered
the emergence of superconsciousness in some individuals as the next step;
in a similar vein the Swiss philosopher Jean Gebser spoke of the coming
of four-dimensional integral consciousness, rising from the prior stages
of archaic, magical, and mythical consciousness. The American mystic
Richard Bucke portrayed cosmic consciousness as the next evolutionary
stage of human consciousness, following the simple consciousness of
animals and the self-consciousness of contemporary humans. Ken Wilber's six-level evolutionary process leads from physical consciousness
pertaining to nonliving matter through biological consciousness associated with animals and mental consciousness characteristic of humans to

subtle consciousness, which is archetypal, transindividual, and intuitive. It leads in turn to causal consciousness and, in the final step, to the ultimate consciousness called Consciousness as Such.

Chris Cowan and Don Beck's colorful theory of spiral dynamics sees contemporary consciousness evolving from the strategic "orange" stage, which is materialistic, consumerist, and success-, image-, status-, and growth-oriented; to the consensual "green" stage of egalitarianism and orientation toward feelings, authenticity, sharing, caring, and community; heading toward the ecological "yellow" stage focused on natural systems, self-organization, multiple realities, and knowledge; and culminating in the holistic "turquoise" stage of collective individualism, cosmic spirituality, and Earth changes.

As already noted in chapter 10, the English engineer C. Maxwell Cade analyzed the states of consciousness manifested by healers in reference to the EEG waves they typically produce. He also analyzed what he considered the high and highest states of consciousness: states of samadhi, satori, or lucid awareness. He identified these states as the fifth state and found that, as with the states of healers, they manifest a moderate amount of beta and theta waves, a wide band of alpha waves, and no waves in the delta region. In yogis, practiced meditators, and psychic individuals the fifth state is remarkably stable. They can maintain this state even while functioning in the everyday context; it appears to have become their natural state of consciousness.

Interestingly and by no means merely coincidentally, a state of consciousness with wide amplitude in the alpha region is known to be the altered state conducive to receiving images and intuitions in the spontaneous, nonsensory mode. It is not surprising that it is prominent in meditators, healers, yogis, shamans, and spiritual people in general.

Cade's fifth state corresponds to Gebser's integral consciousness, Bucke's cosmic consciousness, Wilber's Consciousness as Such, and Cowan and Beck's turquoise state of cosmic spirituality and Earth change. It is the transpersonal stage in the evolution of human consciousness. The physical processes underlying this stage can be understood in reference to the Akashic field (the A-field).

THE ROOTS OF TRANSPERSONAL
CONSCIOUSNESS IN THE A-FIELD

It is standard knowledge that all we experience in our lifetime—all our perceptions, feelings, and thought processes—have cerebral functions associated with them. A-field theory adds that these functions have wave-form equivalents, since our brain, the same as other things in space and time, creates waves in the cosmic plenum. Our wave fronts propagate in the A-field of the plenum and interfere with the wave fronts created by the bodies and brains of other people. The resulting interference patterns are natural holograms. Generations upon generations of humans leave their holographic traces in the A-field. The individual holograms integrate in a superhologram, which is the encompassing hologram of a tribe, community, or culture. The collective holograms interface and integrate in turn with the super-superhologram of all people. This is the collective in-formation pool of humankind.

We can access the information carried by these holograms. On the holographic principle of "like connects with like," we can access first of all the information carried by the hologram of our own brain and body. This is the source of long-term memory, extending back to the womb and even beyond. But our access to A-field holograms is not limited to our own hologram. We can also access the holograms of other people: we can tune our brain to enter into "adaptive resonance" with holograms created by brains and bodies other than our own. As a result we can enter into subtle yet effective contact with different people and with nature. We can even enter into communication with recently deceased people, as the prima facie mind-boggling experience recounted in the annex shows.

PRACTICAL CONSEQUENCES OF THE
QUANTUM SHIFT TO TRANSPERSONAL
CONSCIOUSNESS

Tuning our brain to enter into adaptive resonance with the hologram of other things and other people means moving beyond today's ego-bound and sense organ–limited consciousness to a wider transpersonal consciousness. This shift is likely to have momentous consequences. When people evolve transpersonal consciousness they become aware of their deep ties to each other, to the biosphere, and to the cosmos. They develop greater empathy with people and cultures near and far and greater sensitivity to animals, plants, and the entire biosphere. As a result a new civilization can see the light of day.

The connection between a shift in consciousness and a shift in civilization was envisaged by Native American cultures, including the Maya, Cherokee, Tayta, Xingue, Hopi, Inca, Seneca, Inuit, and Mapuche. They indicate that we are presently living under the Fifth Sun of consciousness and are on the verge of entering the Sixth Sun. The Sixth Sun will bring a new consciousness and with it a fundamental transformation of civilization.

The native cultures were right. Achieving transpersonal consciousness is likely to further progress toward a civilization based on empathy, trust, and solidarity, a Holos-civilization. But will such a civilization come about in time? This we do not know yet. We do know that more and more people will achieve transpersonal consciousness in the coming years, and if we do not destroy our life-supporting environment and decimate our numbers, a critical mass may do so.

The rapid spread of an evolved consciousness is crucial for humanity's future. Whether such consciousness reaches a critical mass will decide if humanity moves in time from the business as usual scenario leading to breakdown to the transformation scenario that leads toward a new civilization. Even if in some societies frustration caused by retrograde politics now catapults people into action, on the global level it is difficult to see how a sufficient number of people would come up with the motivation necessary to achieve a fundamental shift in values, perceptions, and behaviors in the absence of a more evolved consciousness. How could enough people come up with the will to pull together to confront the threats they face in common, to elect leaders who support projects of economic cooperation and intercultural solidarity, to adopt strategies in business where the pursuit of profit and growth is informed by the search for corporate social and ecological responsibility, to bring online an E-Parliament that links parliamentarians worldwide in joint efforts to serve the common good, and to organize an effective network of nongovernmental organizations to restore peace in war-torn regions and ensure an adequate supply of food and water for endangered populations—how could they do all this and more if they do not reach a higher level of consciousness? In the absence of transpersonal consciousness the worldwide motivation needed to take effective action may have to await the actual occurrence of crises and catastrophes—and if these involve major tipping points, shifting to the positive scenario will be difficult, if not impossible.

We need the timely spread of transpersonal consciousness to bring about a shift in civilization. This was recognized already in 1991 by Václav Havel, then president of Czechoslovakia. In his address to a joint session of the U.S. Congress, Havel said, "Without a global revolution in the sphere of human consciousness, nothing will change for the better . . . and the catastrophe towards which this world is headed—the ecological, social, demographic, or general breakdown of civilization—will be unavoidable."

Havel's point was well taken, but it is not a reason for despair: human consciousness *can* evolve. At the innovative frontiers of society it is already evolving This could empower the shift to a new civilization—a civilization of Holos.

This quantum shift in the global brain is humanity's best chance. Margaret Mead said, "Never doubt the power of a small group of people to change the world. Nothing else ever has." Small groups of people with an evolved consciousness will change the world—if they grow into a critical mass in time. There could not be a nobler or more important task in our day than to empower this evolution.

Part Three

GLOBAL*SHIFT*
IN
ACTION
The Club
of Budapest and
Its Initiatives

*The Club of Budapest is an association of ethical and responsible
leading personalities in various parts of the world dedicated to the
proposition that we need urgently to change the world and for that
we must also change ourselves. This part gives a brief account of
the history of the Club, its fields of activity, and the objectives of its
major projects.*

FOURTEEN

A Brief History of the Club of Budapest

Contributed by Iván Vitányi and Mária Sági

The origins of the Club of Budapest can be traced to two events that took place in the second half of the last century. The first was the establishment, in 1968, of the Club of Rome, a progressive think tank with a worldwide base, particularly active in Europe. In Hungary its prominent representative was Ervin Laszlo. The second factor was the particular role played by Hungary during the fall of Communism at the end of the 1980s. Of all the former Soviet bloc countries, it was in Hungary that the transformation to democracy was the most ordered and peaceful, and totally without bloodshed.

There was a vast movement of opinion in Hungary at the time—popular opinion and intellectual opinion, also shared by the progressive echelons of high-ranking politicians—that pointed toward the need for a peaceful transformation from a Soviet satellite to an independent democratic state. Agreement between the parties was negotiated with the result that free elections were held in May of 1990. József Antall, a liberal democrat, was elected Prime Minister. Soviet troop withdrawal commenced immediately thereafter and was completed within a year.

It was the opening of dialogue within the Hungarian political elite that led to the opening of the border with Austria, the historic event that precipitated the dismantling of the Iron Curtain.

Iván Vitányi, the senior author of this short history, was a member of the political-intellectual leadership of the country. As Member of Parliament he took part in negotiating the transfer of power to Hungary's first democratically elected government.

In the course of the 1970s and 1980s, Ervin Laszlo often visited Hungary. The renewal of his long-standing friendship with Vitányi was instrumental in giving birth to the Club of Budapest. There was a surge of interest in the work of the Club of Rome in Hungary following publication of its first report, *The Limits to Growth* (1972). This in turn raised interest in Laszlo's work, especially in the report he headed for the Club of Rome, *Goals for Mankind* (1977). Vitányi in turn became Director of the Institute for Culture in 1970, where Mária Sági was a researcher and later principal collaborator. They worked together on numerous research projects in cultural sociology and social psychology.

Sági and Laszlo started collaborating in 1983. The Institute for Culture held an international conference in December 1983, at which Laszlo gave a lecture on general systems and evolution theory. He was impressed by the institute's work and fascinated by the research of Sági and Vitányi on generative ability in music. He particularly appreciated the application of thorough deep-interview techniques in sociopsychological research, which was not usual among sociologists at that time.

Laszlo became affiliated with the Tokyo-based United Nations University (UNU), and in 1984, when Suzuki Sakura Mushakoji, Vice-Rector of the UNU, was looking for research affiliates in Central Europe, Laszlo recommended the Institute for Culture. An agreement was reached, finances were secured, and the Institute for Culture began its research on "European Identity." Work was completed in eight countries under the direction of Mária Sági, who at the time was the Institute's principal researcher. Sági also headed the research on Hungary, collected the international results, and compiled the final report. The report was published in Hungarian in the periodical *Valóság* and subsequently in English in a special edition of the journal *World Futures*.

In 1984, backed by the Institute for Culture, Vitányi and Laszlo founded the European Culture Impact Research Consortium (EURO-CIRCON). In the years that followed the international research projects of the Institute were carried out under the auspices of EUROCIRCON.

The years between 1988 and 1992 were years of high drama in East-Central Europe. The transition that took place was so fundamental that it would be more correct to call it a transformation. During these critical times Laszlo raised the idea of founding an international "artists' and writers' club" to partner with the Club of Rome. It was to focus in particular on the "soft factors" of the limits to growth: values, expectations, worldviews, and states of mind and consciousness. These, he said, may be even more decisive in our time than money and technology. He suggested that Budapest would provide an ideal intellectual and cultural climate for this enterprise. The idea was taken up by Sándor Csoóri, then President of the World Federation of Hungarians, and the Club of Budapest was called into being. Laszlo was named President and was supported by a Board made up of Sándor Csoóri (poet), Sándor Sára (film director), Gedeon Dienes (dance historian), and Mária Sági and Iván Vitányi (cultural sociologists). The Club was given office space in the House of Hungarian Culture, where it remains to this day.

The Club got off to a slow start as its first General Secretary worked in tourism and did not devote sufficient time to Club activities. It was only in 1995 that real work began. By the following year preparations for the first conferences were well in hand, some two dozen world-famous personalities had joined the Club as Honorary Members, and the Club published the *Manifesto on Planetary Consciousness*. This document states the fundamental objectives and enduring mission of the Club of Budapest and deserves to be reproduced in full, as it is in the following chapter.

THE CLUB OF BUDAPEST

Headquarters:
The Club of Budapest Foundation (Hungary)

Ervin Laszlo, President; Mária Sági, Scientific Secretary
International Coordination Office: The Global*Shift* Network (Germany)
Johannes Heimrath and Wolfgang Riehn, Directors
www.clubofbudapest.org

Honorary Members

Dsingis AITMATOV
writer

Oscar ARIAS
statesman, Nobel Peace laureate

Dr. A.T. ARIYARATNE
Buddhist spiritual leader

Maurice BÉJART
dancer/choreographer

Dr. Thomas BERRY
scientist/Christian spiritual leader

Sri BHAGAVAN
Indian spiritual leader

Sir Arthur C. CLARKE
writer

Paolo COELHO
writer

H.H. The XIVth DALAI LAMA
statesman/spiritual leader

Dr. Riane EISLER
feminist historian/activist

Milos FORMAN
film director

Peter GABRIEL
musician

Hans-Dietrich GENSCHER
statesman

Dr. Jane GOODALL
scientist/youth activist

Rivka GOLANI
musician

Mikhail GORBACHEV
political leader

Arpád GÖNCZ
writer/statesman

Václav HAVEL
writer/statesman

Dr. Hazel HENDERSON
economist/activist

Bianca JAGGER
activist

Miklós JANCSÓ
film director

Ken-Ichiro KOBAYASHI
orchestra director

THE CLUB OF BUDAPEST (CONT.)

Gidon KREMER
musician

Dr. Hans KÜNG
theologian/Christian spiritual
  leader

Dr. Shu-hsien LIU
philosopher

Zubin MEHTA
orchestra director

Lord Yehudi MENUHIN+
musician

Dr. Edgar MITCHELL
scientist/astronaut

Dr. Edgar MORIN
philosopher/sociologist

Dr. Robert MULLER
educator/activist

Gillo PONTECORVO
film director

Jean-Pierre RAMPAL+
musician

Mary ROBINSON
political and human rights leader

Mstislav ROSTROPOVICH+
orchestra director

Sir Josef ROTBLAT+
scientist/activist/Nobel Peace
  laureate

Dr. Peter RUSSELL
philosopher/futurist

Masami SAIONJI
Japanese spiritual leader

Dr. Karan SINGH
statesman/Hindu spiritual leader

Sir George SOLTI+
orchestra director

Rita SÜSSMUTH
political leader

Sir Sigmund STERNBERG
Jewish spiritual leader

His Grace Desmond TUTU
Christian spiritual leader

Liv ULLMANN
film actor/director

Sir Peter USTINOV+
actor/writer/director

Vigdis FINNBOGADOTTIR
political leader

Richard von WEIZSÄCKER
statesman

Dr. Elie WIESEL
writer/Nobel Peace laureate

Betty WILLIAMS
activist/Nobel Peace laureate

Dr. Mohammed YUNUS
economist/financial leader

+ deceased

FIFTEEN

Manifesto on Planetary Consciousness

*Drafted by Ervin Laszlo in collaboration with
the Dalai Lama and adopted by the
Club of Budapest on October 26, 1996*

THE NEW REQUIREMENTS OF THOUGHT
AND ACTION

1. In the closing years of the twentieth century, we have reached a crucial juncture in our history. We are on the threshold of a new stage of social, spiritual, and cultural evolution, a stage that is as different from the stage of the earlier decades of this century as the grasslands were from the caves, and settled villages from life in nomadic tribes. We are evolving out of the nationally based industrial societies that were created at the dawn of the first industrial revolution and heading toward an interconnected, information-based social, economic, and cultural system that straddles the globe. The path of this evolution is not smooth: it is filled with shocks and surprises. This century has witnessed several major shock waves, and others may come our way before long. The way we shall cope with present and future shocks will decide our future and the future of our children and grandchildren.

2. The challenge we now face is the challenge of choosing our destiny. Our generation, of all the thousands of generations before us, is called upon to decide the fate of life on this planet. The processes we have initiated within our lifetimes and the lifetimes of our parents and grandparents cannot continue in the lifetimes of our children and grandchildren. Whatever we do will either create the framework for reaching a peaceful and cooperative global society and thus continuing the grand adventure of life, spirit, and consciousness on Earth or set the stage for the termination of humanity's tenure on this planet.

3. The patterns of action in today's world are not encouraging. Millions of people are without work; millions are exploited by poor wages; millions are forced into helplessness and poverty. The gap between rich and poor nations, and between rich and poor people within nations, is great and still growing. Though the world community is relieved of the specter of superpower confrontation and is threatened by ecological collapse, the world's governments still spend a thousand billion dollars a year on arms and the military and only a tiny fraction of this sum on maintaining a livable environment.

4. The militarization problem, the developmental problem, the ecological problem, the population problem, and the many problems of energy and raw materials will not be overcome merely by reducing the number of already useless nuclear warheads nor by signing politically softened treaties on world trade, global warming, biological diversity, and sustainable development. More is required today than piecemeal action and short-term problem solving. We need to perceive the problems in their complex totality and grasp them, not only with our reason and intellect, but also with all the faculties of our insight and empathy. Beyond the powers of the rational mind, the remarkable faculties of the human spirit embrace the power of love, of compassion, and of solidarity. We must not fail to call upon their remarkable powers when confronting the task of initiating the embracing, multifaceted approaches that alone could enable us to reach the next stage in the evolution of our sophisticated but unstable and vulnerable sociotechnological communities.

5. If we maintain obsolete values and beliefs, a fragmented consciousness, and a self-centered spirit, we also maintain outdated goals and behaviors. And such behaviors by a large number of people will block the entire transition to an interdependent yet peaceful and cooperative global society. There is now both a moral and a practical obligation for each of us to look beyond the surface of events, beyond the plots and polemics of practical policies, the sensationalistic headlines of the mass media, and the fads and fashions of changing lifestyles and styles of work, an obligation to feel the groundswell underneath the events and perceive the direction they are taking: to evolve the spirit and the consciousness that could enable us to perceive the problems and the opportunities—and to act on them.

A CALL FOR CREATIVITY AND DIVERSITY

6. A new way of thinking has become the necessary condition for responsible living and acting. Evolving it means fostering creativity in all people, in all parts of the world. Creativity is not a genetic but a cultural endowment of human beings. Culture and society change fast, while genes change slowly: no more than one half of one percent of the human genetic endowment is likely to alter in an entire century. Hence most of our genes date from the Stone Age or before; they could help us to live in the jungles of nature but not in the jungles of civilization. Today's economic, social, and technological environment is our own creation, and only the creativity of our mind—our culture, spirit, and consciousness—could enable us to cope with it. Genuine creativity does not remain paralyzed when faced with unusual and unexpected problems; it confronts them openly, without prejudice. Cultivating it is a precondition of finding our way toward a globally interconnected society in which individuals, enterprises, states, and the whole family of peoples and nations could live together peacefully, cooperatively, and with mutual benefit.

7. Sustained diversity is another requirement of our age. Diversity is basic to all things in nature and in art: a symphony cannot be made of one tone or even played by one instrument; a painting must have many shapes and perhaps many colors; a garden is more beautiful if it contains

flowers and plants of many different kinds. A multicellular organism cannot survive if it is reduced to one kind of cell; even sponges evolve cells with specialized functions. And more complex organisms have cells and organs of a great variety, with a great variety of mutually complementary and exquisitely coordinated functions. Cultural and spiritual diversity in the human world is just as essential as diversity in nature and in art. A human community must have members that are different from one another not only in age and sex but also in personality, color, and creed. Only then could its members perform the tasks that each does best and complement each other so that the whole formed by them could grow and evolve. The evolving global society would have great diversity, were it not for the unwanted and undesirable uniformity introduced through the domination of a handful of cultures and societies. Just as the diversity of nature is threatened by cultivating only one or a few varieties of crops and husbanding only a handful of species of animals, so the diversity of today's world is endangered by the domination of one, or at the most a few, varieties of cultures and civilizations.

8. The world of the twenty-first century will be viable only if it maintains essential elements of the diversity that has always hallmarked cultures, creeds, and economic, social, and political orders as well as ways of life. Sustaining diversity does not mean isolating peoples and cultures from one another: it calls for international and intercultural contact and communication with due respect for each other's differences, beliefs, lifestyles, and ambitions. Sustaining diversity also does not mean preserving inequality, for equality does not reside in uniformity but in the recognition of the equal value and dignity of all peoples and cultures. Creating a diverse yet equitable and intercommunicating world calls for more than just paying lip service to equality and just tolerating each other's differences. Letting others be what they want as long as they stay in their corner of the world and letting them do what they want "as long as they don't do it in my backyard" are well meaning but inadequate attitudes. As the diverse organs in a body, diverse peoples and cultures need to work together to maintain the whole system in which they are a part, a system that is the human community in its planetary abode. In the last decade of the twen-

tieth century, different nations and cultures must develop the compassion and the solidarity that could enable all of us to go beyond the stance of passive tolerance, to actively work with and complement each other.

A CALL FOR RESPONSIBILITY

9. In the course of the twentieth century, people in many parts of the world have become conscious of their rights as well as of many persistent violations of them. This development is important, but in itself it is not enough. In the remaining years of this century we must also become conscious of the factor without which neither rights nor other values can be effectively safeguarded: our individual and collective responsibilities. We are not likely to grow into a peaceful and cooperative human family unless we become responsible social, economic, political, and cultural actors.

10. We human beings need more than food, water, and shelter; more even than remunerated work, self-esteem, and social acceptance. We also need something to live for: an ideal to achieve, a responsibility to accept. Since we are aware of the consequences of our actions, we can and must accept responsibility for them. Such responsibility goes deeper than many of us may think. In today's world all people, no matter where they live and what they do, have become responsible for their actions as:

- private individuals
- citizens of a country
- collaborators in business and the economy
- members of the human community
- persons endowed with mind and consciousness

As individuals, we are responsible for seeking our interests in harmony with, and not at the expense of, the interests and well-being of others; responsible for condemning and averting any form of killing and brutality; responsible for not bringing more children into the world than we truly need and can support; and responsible for respecting the right to life, development, and equal status and dignity of all the children, women, and men who inhabit the Earth.

As citizens of our country, we are responsible for demanding that our leaders beat swords into plowshares and relate to other nations peacefully and in a spirit of cooperation, that they recognize the legitimate aspirations of all communities in the human family, and that they do not abuse sovereign powers to manipulate people and the environment for shortsighted and selfish ends.

As collaborators in business and actors in the economy, we are responsible for ensuring that corporate objectives do not center uniquely on profit and growth but include a concern that products and services respond to human needs and demands without harming people and impairing nature, do not serve destructive ends and unscrupulous designs, and respect the rights of all entrepreneurs and enterprises who compete fairly in the global marketplace.

As members of the human community, it is our responsibility to adopt a culture of nonviolence, solidarity, and economic, political, and social equality, to promote mutual understanding and respect among people and nations whether they are like us or different, and to demand that all people everywhere should be empowered to respond to the challenges that they face with the material as well as spiritual resources that are required for this unprecedented task.

And as persons endowed with mind and consciousness, our responsibility is to encourage comprehension and appreciation for the excellence of the human spirit in all its manifestations and for inspiring awe and wonder for a cosmos that brought forth life and consciousness and holds out the possibility of its continued evolution toward ever higher levels of insight, understanding, love, and compassion.

A CALL FOR PLANETARY CONSCIOUSNESS

11. In most parts of the world, the real potential of human beings is sadly underdeveloped. The way children are raised depresses their faculties for learning and creativity; the way young people experience the struggle for material survival results in frustration and resentment. In adults this leads to a variety of compensatory, addictive, and compulsive behaviors.

The result is the persistence of social and political oppression, economic warfare, cultural intolerance, crime, and disregard for the environment. Eliminating social and economic ills and frustrations calls for considerable socioeconomic development, and that is not possible without better education, information, and communication. These, however, are blocked by the absence of socioeconomic development, so that a vicious cycle is produced: underdevelopment creates frustration, and frustration, giving rise to defective behaviors, blocks development. This cycle must be broken at its point of greatest flexibility, and that is the development of the spirit and consciousness of human beings. Achieving this objective does not preempt the need for socioeconomic development with all its financial and technical resources but calls for a parallel mission in the spiritual field. Unless people's spirits and consciousnesses evolve to the planetary dimension, the processes that stress the globalized society/nature system will intensify and create a shock wave that could jeopardize the entire transition toward a peaceful and cooperative global society. This would be a setback for humanity and a danger for everyone. Evolving human spirit and consciousness is the first vital cause shared by the whole of the human family.

12. In our world static stability is an illusion; the only permanence is in sustainable change and transformation. There is a constant need to guide the evolution of our societies so as to avoid breakdowns and progress toward a world where all people can live in peace, freedom, and dignity. Such guidance does not come from teachers and schools, not even from political and business leaders, though their commitment and roles are important. Essentially and crucially, it comes from each person himself and herself. An individual endowed with planetary consciousness recognizes his or her role in the evolutionary process and acts responsibly in light of this perception. Each of us must start with ourselves to evolve our consciousness to this planetary dimension; only then can we become responsible and effective agents of our society's change and transformation. Planetary consciousness is the knowing as well as the feeling of the vital interdependence and essential oneness of humankind and the conscious adoption of the ethics and the ethos that this entails. Its evolution is the basic imperative of human survival on this planet.

SIXTEEN

Principal Activities of the Club of Budapest

The Club of Budapest:

1. Collaborates in the development and operation of a global degree-granting institution of higher learning with the vital mission of educating and training a new generation of positive change agents able to work in existing and newly created public and private institutions toward the solution of the world's most vexing and urgent problems (the Global*Shift* University).
2. Participates in an alliance it has cofounded to elucidate the concept of a new civilization and the practical paths leading to it, together with like-minded institutes and organizations (the Creating a New Civilization Alliance).
3. Participates in a further alliance it has likewise founded to help advanced-thinking individuals and organizations exchange information, adopt shared goals and objectives, and join together in the implementation of their activities and projects (the World Wisdom Alliance).
4. Brings together leading thinkers and visionaries to discuss the problems facing the world community and produce a realistic vision of a civilization that is peaceful and sustainable, in balance with itself and with nature (the World Wisdom Council).

5. Conducts periodic surveys of the new cultures of solidarity and responsibility in various countries and regions of the world to establish their size, composition, and values and to communicate the results to the new cultures as well as to the public at large (the International Survey of Emergent Cultures).

6. Organizes a series of global events to support and celebrate the emerging culture of peace and solidarity on Earth (the Global Peace Meditation/Prayer Days).

7. Offers annual prizes to individuals whose thinking and activity exemplify the new culture and the new consciousness (the Planetary Consciousness Prizes) and confers awards on organizations and communities whose activities and projects translate the new consciousness into action (the Best Practice Awards).

8. Produces a series of books and reports and related visual materials that describe the critical condition of the human community, trace the roots of the present unsustainability, and apply the latest insights of the sciences to offer practical guidelines for evolving a planetary civilization (the *You Can Change the World* book series).

SEVENTEEN

Objectives of the
Global*Shift* University

The initiative to create the Global*Shift* University (GSU) is grounded in
the belief that a fundamental *shift* is required in order to create a culture
of sustainability. Current global challenges demand concerted action by
all conscious individuals—informed and effective action that is optimized
for maximum constructive impact. A growing number of concerned persons and institutions on every continent are currently involved in addressing these challenges, but there are numerous barriers that impair their
ability to act on their vision and motivation. The required marshaling
and channeling of common interest–based, planet-wide activism cannot
be expected to occur spontaneously or evolve organically as long as social
cohesion is driven by national ethnocentrism, individual egocentricity, and
the special interests of public and private institutions. Four Fundamental
Elements are needed to overcome this societal fragmentation.

- The First Fundamental Element is an overarching vision of a positive future, the proverbial "big ideas" that capture and galvanize
  the imagination of people. While there is no shortage of visionaries and big thinkers, there is no widely recognized and embraced
  framework or architecture for effective cultural, social, political,
  and economic reorganization. The lack of a systemic, commonly

held, values-driven strategy for making positive change, grounded in science, spirituality, and fairness, diminishes and undermines the possibility of effective collective action.

- The Second Fundamental Element is a set of effective communication processes and protocols that pierce the barrier of fragmentation, letting in the light of a new vision and conveying big ideas and interconnecting those who are ready to receive them. This would create synergisms and forward momentum toward planetary sustainability and peace among individuals, public institutions, and civil society as well as business organizations.
- The Third Fundamental Element regards the key skills of interdependent action, today all too rarely present, such as the ability to develop covenantal relationships, engage in meaningful dialogue, engage in critical thinking, develop collective wisdom, and translate vision and ideas into meaningful action and robust processes of change.
- The Fourth Fundamental Element is a new perception of the role that timely and effective engagement by both business and government can play in collective action and the collaborative promotion of global sustainability.

INSTITUTIONAL MISSION AND GOALS

It is the mission and goal of GSU to be a key participant in the assembling and dissemination of the here cited Four Fundamental Elements in cooperation, collaboration, and coordination with individuals and organizations dedicated to the pursuit of the same and related ends. The University is dedicated to:

1. Addressing timely problems, promoting instantaneous acquisition of relevant information, facilitating development of insight and the pursuit of possibility, and enabling learning through direct experience and formal study that leads to the acquisition of academic degrees (baccalaureate and masters).

2. Promoting personal and organizational transformation that enables graduates to serve as agents for positive change as they work to create values and visions that enable lasting world peace.

3. Serving as a coalescing agent for social action and cultural awareness gained through learning, research, and authentic application.

4. Advancing the value of living in accordance with the cyclical processes of nature and fostering compliance with their evolutionary laws.

5. Championing and facilitating effective actions to transform a conflict-riddled world to one of sustainability and peace.

The overarching goal of GSU is to support the development, transmission, and use of knowledge to promote ethical and ecologically responsible behavior. To this end, GSU will carry out its mandate in accordance with five interconnected and mutually reinforcing purposes:

- To increase public understanding of the nature and scope of sustaining local, regional, national, and global ecologies, as well as of approaches to their management, through degree and nondegree academic programs, print and web media publications, and electronic forms and forums of communication.

- To expand society's knowledge regarding the changing nature of the global environment, cultural, political, social, and economic value systems, and paths to inner growth and enlightenment, identifying "best practices" in an interdependent global system where the actions of each affect the destiny of all.

- To support policymakers in devising and implementing holistic strategies by mobilizing scholars and providing independent creative assessments of how to approach and resolve issues concerned with enhancing the sustainability of local and global environments.

- To facilitate the resolution of issues of ethics, social responsibility, and global coordination by organizing and supporting transformative dialogues within and among parties of political, economic, cultural, religious, scientific, and social influence, and providing technical assistance for addressing emerging issues of sustainability and ethics.

- To strengthen the education of successive generations by building the capacity of institutions from high school through graduate study to better address issues of planetary consciousness and social and ecological sustainability.

EIGHTEEN

Objectives of the World Wisdom Council

In recognition of today's accelerating global challenges and opportunities in all areas of human concern and the need to address them in a wise and considered manner free of political, economic, and ideological constraints, the Club of Budapest, in association with the World Commission on Global Consciousness and Spirituality, created the "World Wisdom Council" (WWC). The WWC is an independent body, with its initial members drawn from the Club of Budapest and the World Commission.

The WWC is to represent the collective wisdom of humanity as a whole, both masculine and feminine, from every continent, major culture, and religion. Its core mission is to transcend narrow national agendas and self-serving individual interests, recognizing that thinking based on these levels cannot meet today's growing global challenges.

The WWC has been convened in the conviction that the paramount requirement in this age of discontinuity and transformation is to recognize that, through the development of a new dimension of consciousness, the world can be constructively changed by women and men wherever they live and whatever their interests and lot in life. It is to have a unique role in helping today's people and societies find a feasible path toward a more peaceful and sustainable future.

TASK AND MISSION

The task of the Council is to build on the power and creativity innate in all people by:

1. Bringing to the attention of the widest layers of the public both the dangers and the opportunities inherent in the human condition in its global dimension.
2. Identifying priority areas where individual and cooperative action is needed in order to reinforce progress toward peace and sustainability, locally as well as globally.
3. Offering guidance for developing the individual and collective wisdom that empowers action capable of bringing about constructive change in the local and the global economic, social, and ecological environment.

The World Wisdom Council is politically, socially, and culturally nonpartisan, championing the joint interest of all humans and all life on this planet and informing people so that they can move toward a world where they can live in peace with each other and in harmony with nature. Its members realize that there is already a growing range of initiatives aligned with the WWC mission. In consequence it is taking as one of its highest priorities the formation of networks, partnerships, and collaborations in the interest of mobilizing the forces required for constructive transformation on a global scale.

NINETEEN

Objectives of the International Survey of Emergent Cultures

The International Survey of Emergent Cultures is based on three inter-related considerations.

- *The values and the ethics that shape today's world produce socio-economic and ecological trends that are not sustainable.* The continued dominance of this cultural paradigm would lead to crisis conditions that are not in the best interest of any segment of the international community.
- *The changes occurring in the values and the ethics of a significant segment of the population in countries East and West, North and South can create a basis for moving toward greater socioeconomic and ecological sustainability.* Although these changes are occurring, they are not generally known, and consequently the people who manifest them lack the economic and political weight to affect conditions in the local and the global environment.
- *Providing information that leads to the fuller recognition of current changes in values and ethics would enable the emerging cultures to achieve higher visibility and greater societal weight.*

Public behaviors and aspirations leading to a positive modification of currently unsustainable socioeconomic and ecological trends would be facilitated by the realization that the emerging cultures (the "cultural creatives" or their equivalent) are widespread and encompass diverse strata in contemporary societies.

The evolution of the values and the ethics of people in all walks of life and parts of the world is the best and most reliable avenue toward the creation of a more peaceful and sustainable world. This evolution is occurring already, but it is not widely recognized. A better understanding of its dimensions would have important consequences for every country and continent. Those who partake in this development would be reinforced by the knowledge that they are not alone. Those who are not partaking in it would be encouraged to reexamine their values, their ethics, and their behavior. The business community would find that there is an emerging market that is not fully recognized and hence not adequately served. And the political community would become aware that there is a potentially important segment of their constituency whose interests and aspirations do not yet have political representation. The consequences would include changes in the general culture of the public, a shift in business strategies, and a reshaping of the current political landscape.

The International Survey of Emergent Cultures is to:

- Document the existence of emerging cultures of responsible thinking and acting in different parts of the world.
- Research the nature and the dimensions of these cultures.
- Make available the findings to those who are part of the new cultures as well as to people at large.

TWENTY

Objectives of the Global Peace Meditation/Prayer Days

The first Global Peace Meditation/Prayer Day was created to reduce the level of conflict and violence in the world and to help create deeper understanding, tolerance, and readiness to live in peace with our neighbors both near and far, as well as with nature. On Sunday, May 20, 2007, an estimated one million people participated in the first Global Peace Meditation/Prayer Day in sixty-five countries on the five continents. Never before have so many people in so many countries and from so many faiths and cultures come together to direct the power of their meditation and prayer to peace on Earth: the first truly common cause of all of humanity.

Numerous tests and experiments have shown that deep prayer and meditation can heal people, heal other species, and create peace and harmony in human communities. Now for the first time the power of prayer and meditation has been directed at the entire community of humans on the planet, with over a million entering a deeper state of consciousness and giving expression to their heartfelt wish that "peace may prevail on Earth."

The collaboratively organized meditations of May 20, 2007, followed the same procedure wherever they took place and regardless of the culture, faith, and religion of the participants. The events began with initial speeches, music, and dance and were followed by meditation or prayer guided by a spiritual master. They ended with five minutes of silence when the participants stood and held hands and then silently repeated a phrase such as "May peace prevail on Earth."

The one hour meditations/prayers were carefully synchronized to reinforce each other and produce the maximum effect. The first group of events took place at the same time in eastern and western Australia and in Japan. The second group brought together people in India, central Africa and South Africa, Israel, Greece, Hungary, Germany, Italy, and England. The third group embraced Canada, the United States, Venezuela, Brazil, Argentina, Uruguay, Chile, as well as Hawaii and Samoa.

The network of groups that registered for the Global Day emerged rapidly, as the website www.globalpeacemeditationprayerday.org announced the event, listed the participating groups, and invited others to join.

Building on the promise of the 2007 Global Peace Meditation/Prayer Day, the Club of Budapest, in partnership with the World Prayer Society, will continue to create periodic Global Peace Meditation/Prayer Days to bring together not just one but many millions of dedicated people and focus the power of their consciousness on peace in the world. Such a "critical mass" of humans will, the Club of Budapest believes, make a major and possibly crucial contribution toward achieving a world that is peaceful, humane, and sustainable.

Communication Beyond the Grave

EXPLORING AN EXPLANATION

Evidence coming to light at the frontiers of science shows that on a deep level all things in the cosmos are connected with all other things. This connection, as we have seen, holds true for human brains: in altered states of brain and mind the brain waves of different individuals, even if separated by finite distances, become synchronized and the rhythms exhibited by one are picked up by the others. The universe, including human beings, is fundamentally and, as it appears, "nonlocally" coherent.

The surprising nonlocal coherence of the brains and minds of different individuals produces a number of so-called paranormal phenomena including telepathy, remote viewing, remote healing, twin pain, and distance-independent connection among emotionally bonded people. In these cases the nonlocal coherence between the minds of individuals is mediated by their brains and bodies.

But what about coherence and connection with a mind that is no longer associated with a living brain and body? The suggestion that we can be "in touch" with people who have already died is mind-boggling. Yet numerous experiences and experiments in the growing field of ITC (instrumental transcommunication) testify that communication beyond the grave may exist. The evidence is there, but a satisfactory explanation for it is still lacking.

This annex takes on the intriguing and—from a deep human perspective—extraordinarily important task of looking for a scientific explanation for ITC. It does so by reviewing an experience this writer had recently and exploring an explanation of it in light of Akashic field theory.

THE EXPERIENCE

April 7, 2007. I am sitting in a darkened room in the Italian town of Grosseto, together with a group of sixty-two other people. It is evening, and there is not a sound, other than the sounds of the shortwave band of a radio. It is an ancient valve radio, the kind that works not with transistors but with vacuum tubes. I am sitting on a small stool immediately behind an old Italian who wears a hat and is dressed as if it were still winter although it is warm in the room—and getting warmer by the minute.

The Italian—a renowned psychic who considers himself not a commercial medium but a serious psychic researcher—is Marcello Bacci. For the past forty years he has been hearing voices through his radio and has become convinced that they are the voices of people who have passed away. Those who come to his regular "dialogues with the dead" are likewise convinced of this. They are people who have lost a son or a daughter, a father or a mother or a spouse, and hope to have the experience of hearing them talk through Bacci's radio.

We have been sitting in the darkened room for a full hour. Bacci is touching the wooden box that houses the radio with both hands, caressing it on the sides, at the bottom and on the top, and speaking to it. "Friends, come, speak to me, don't hesitate, we are here, waiting for you . . ." But nothing happens. As Bacci plays with the dial, the radio emits either the typical shortwave static or conveys one or another shortwave broadcast. I am getting convinced that the doubts I had initially entertained were justified: after all, how could a shortwave receiver pick up voices from the "other side"? How could the "other side" transmit signals through the electromagnetic spectrum? A far-out supposition! But everyone is waiting. Bacci keeps caressing the radio, turning the dial, and asking for the voices. I sit behind him, and wait for a miracle . . .

And then: there are sounds like heavy breathing or like a rubber tube or pillow pumped with air. Bacci says, "At last!" He continues to move the dial, but there are no longer any shortwave transmissions coming through. Wherever he turns the dial, the radio transmits only the periodic breathing. The entire radio seems to have become tuned to this one frequency, one that an associate of Bacci is carefully monitoring on a device to my right.

Bacci talks to the radio, encouraging whoever or whatever is breathing, or pumping air, to talk back to him. Now voices are coming through on the air. Indistinct, hardly human voices, difficult to understand, but they speak Italian, and Bacci seems to understand. The entire room freezes in concentration. The first voice is that of a man. Bacci talks to him, and the voice answers. Bacci tells him that there are many people here tonight (the usual group is no more than twelve), and they are all anxious to get into a conversation.

Bacci says that behind him—immediately to my left—sits someone whom the voice knows. "Who is he?" (He is the renowned French psychic-researcher Father Brune, who has written several books on his experiences of talking with deceased people. He lost his brother about a year ago and has contacted him since and hopes to do so again.) The voice answers, "Père Brune" (as Fr. Brune is known in his native France). Fr. Brune asks, "With whom am I speaking?" It turns out that it is not his brother but Father Ernetti, a close friend and associate of Fr. Brune who died not long ago.

(I subsequently found out that Fr. Ernetti, a Roman Catholic priest attached to the Abbey of San Giorgio Maggiore in Venice, began having instrumental transcommunications in 1952. Together with Fr. Gemelli, an eminent medical doctor at the Catholic University of Milan, he had been investigating ways of filtering the audiotapes of Gregorian chants to enhance the purity of their sound. They were frustrated by the fact that the wire used by the ancient recorders broke frequently and required constant and delicate repair. Finally Father Gemelli, as was his habit when exasperated, called on his deceased father for help. When they restarted their own recorder the two fathers heard the voice of Gemelli senior rather than the Gregorian chant on which they were working. It said, "Of course I'll help you! I am always with you." The fathers reported the incident to Pope Pious XII who gave them a highly positive response: Hearing the voice could initiate "a new scientific study for confirming faith in the afterlife." Fr. Brune was privy to these facts and became a long-lasting friend and co-researcher of Fr. Ernetti.)

Fr. Brune and Fr. Ernetti talk for a while, and then Bacci—who continues to lean forward and caress the radio—says, "Do you know who else is sitting here, just behind me?" A voice that seems to be different, but also male, says "Ervin." He pronounces it as one does in Hungarian or in German, with the "E" as in "extraordinary" and not, as in English, as in "earth." Bacci asks, "Do you know who he is?" and the voice answers, "É ungherese" (He is Hungarian). The voice then gives my family name but pronounces it as Italians sometimes do: "Latzlo," and not as Hungarians, with a soft "s" as in "Lasslo."

Bacci asks for my hand—I am sitting immediately behind him—and places my hand on his. His wife and long-standing associate places her hand on mine. My hand is sandwiched between theirs and is getting warmer—indeed, quite hot. Bacci tells me, "Speak to them in Hungarian." I lean forward and do so. My voice is choked, for I am moved. The unthinkable is happening, just as I hoped but hardly dared to expect. I say how happy I am to speak with them. I do not think I should ask whether they are dead (how do you say to someone you talk to, "Are you dead?") but ask instead, "Who are you, and how many are you?" The answer that comes in Hungarian is indistinct but I can make it out: "We are all here" (a voice adds: "The Holy Spirit knows all languages"). Then I ask, "Is it difficult for you to talk to me like this?" (thinking of the seemingly strenuous breathing that preceded the conversation). A woman answers, quite clearly, and in Hungarian: "We have some difficulties (or obstacles), but how is it for you, do you have obstacles too?"

I say, "It was not easy for me to find this way of talking with you, but now I could do it and I am delighted."

Bacci is thinking of the many people who hope to have contact with their lost loved ones and directs attention to the others in the room, not identifying anyone by name, just recalling that they, too, would like to get answers. The voice—the same or a different male voice, it is difficult to say for certain—comes up with a number of names, one after the other. The person named speaks up, often in a voice trembling with hope. "Can I hear Maria (or Giovanni . . .)?" Sometimes a younger voice comes on the air, and a person in the room gives a shout of delight and recognition.

And so it continues for about half an hour. There are breaks taken up by the sound of air rushing as in heavy breathing (Bacci explains, "They are recharging themselves"), but the voices come back. Then it appears that they are really gone. Bacci moves the dial on the shortwave band, but only static and some shortwave broadcasts come through, as they did during the first hour. He gets up, the lights are switched on; the séance is over.

Everything has been carefully recorded, both on audio and on film (a professional film crew has been working silently in the dark, but not entirely pitch-dark, room). It is time to ask: What has happened? What am I to make of it?

THE QUESTION OF AUTHENTICITY

Could it be that the above experience in instrumental transcommunication was a hoax? Could there have been devices hidden in the room or at distant locations, connected electronically with the radio, which had produced the sounds I had heard?

This possibility cannot be absolutely ruled out, but the record weighs against it. Bacci has conducted these experiments for nearly forty years and during this time they have been witnessed by scientists as well as electronic engineers. The most exhaustive tests were made in 1996 by Prof. Mario Festa, a nuclear physicist at the University of Naples. In the presence of researchers from Il Laboratorio, an Italian research center dedicated to the investigation of the authenticity of paranormal voice phenomena, Festa tested the corresponding electric and magnetic fields while the voices were being heard. Beforehand, with the radio off, the electric field measured 0.71 V/m (volts per meter) and the magnetic field measured 0 mT (millitesla). When the radio was switched on, the electric field rose to 2.15 V/m and the magnetic field to 0.11 mT. During the time the voices manifested themselves the electric field oscillated between 0.54 and 0.81 V/m, and the magnetic field remained at 0 mT. This was surprising and clearly anomalous. A transmission from a normal source—whether a regular sender or an attempted hoax—would have raised the fields to the values associated with the radio's usual operation.

Not satisfied with this evidence, Festa, together with electronic engineer Franco Santi, removed both the frequency modulation valve and the local intermediate oscillation valve from Bacci's radio while the voices were being heard. This would have silenced radio transmission across all wave bands. Yet the voices continued unaltered, without noticeable loss of signal. They continued also when Festa moved the tuning up and down the frequencies. (Bacci did the same when I was present.) In his report, published in 2002, Festa concluded that the results confound the known laws of physics.

(A curious fact about the timing of the experiment I attended is worthy of note. It started, as it had in previous months, precisely at 7:30 p.m. But the voices started only one hour later, when our watches showed 8:30 p.m. Not long beforehand Europe had shifted from winter to summer time. Hence 8:30 p.m. had previously been 7:30 p.m., the exact time the voices manifested themselves. The voices were on time; it was Bacci who attempted communication too soon. Contact, it seems, depends more on the time associated with the rotation of the planet than with the setting of our watches.)

In light of our accepted knowledge of the physical world and its phenomena, Bacci's experiment is clearly anomalous. But it is far from unique. ITC is a repeatedly investigated phenomenon. A number of books and articles have been devoted to recounting the experiments as well as the measures designed to test their authenticity.

RELATED EXPERIMENTS

ITC, the same as the wider body of experimentation known collectively as EVP (electronic voice phenomena), requires an electronic instrument, hence it dates mainly from the 1960s. The pioneer in this field was Dr. Konstantin Raudive, whose classic book *Breakthrough* was published in 1971. Raudive recorded some 72,000 voices emitted by unexplained and seemingly paranormal sources, of which 25,000 contained identifiable words. Since then a wide range of controlled experiments have been carried out; here merely a brief sampling will suffice (David Fontana provides a good overview in his book *Is There an Afterlife?*, chapter 14).

Hans Otto König, an electroacoustics engineer in Germany, experimented with various sources of background noise, including running water in addition to radio static. He noticed that all these carried sounds that reach into the ultrasonic range, whereas regular tape recorders do not register sounds above 20,000 Hertz. He then designed a source of background sound consisting of four sound generators that produce a complex mixture of frequencies above the audible range of human hearing. This worked. König received anomalous voice communications over a period of years.

As news of this spread, he was invited to give a live demonstration on Radio Luxembourg, a popular radio station heard over much of Europe. Independent electrical engineers tested the equipment and monitored the experiment that was witnessed on the air by thousands. The audio equipment designed by König was not operated by him but by independent technicians. When a technician asked that the anomalous communicators make themselves heard, a clear voice answered "Otto König makes wireless with the dead." The reply to a further question was equally clear: "We hear your voice." At the conclusion of the broadcast, the program's presenter, Rainer Holbe, a well-known master of ceremonies, reported in a shaking voice, "I swear by the life of my children that nothing has been manipulated. There are no tricks. It is a voice and we do not know from where it comes." (The broadcast experiment was reported in full by J. G. Fuller in his book *The Ghost of 29 Megacycles*.)

Another series of remarkable experiments was carried out by Dr. Anabela Cardoso, a senior diplomat from Portugal. She first used foreign language broadcasts for background noise but then switched to white noise, using old-fashioned valve radios. Having received answers to her questions both on tape and directly through the radio, she became convinced of the authenticity of the phenomenon and the need for its further study. She established an international publication, *The ITC Journal*, that carries research reports in Portuguese, Spanish, and English. Her own communicators spoke mostly Portuguese, with occasional communications in German, Spanish, and English, all languages in which Dr. Cardoso was fluent. According to David Fontana, who witnessed several of Cardoso's experiments, the possibility of fraud or interference by other persons can be effectively ruled out.

The various forms of ITC include radios, TVs, telephones, computers, and other technical devices. Mark Macy used a device called "the luminator" in combination with an off-the-shelf Polaroid 600 camera and stock film and has obtained thousands of pictures of "spirit faces" that appear on the film in his presence and sometimes also in the presence of a person to whom a given spirit face was known. (The "luminator" was invented by Patrick Richards of Michigan. It has two counterrotating fans that pull air into vents at the base of the unit and blow it out through the vents at the top. The air passes through a Plexiglas barrel lined with rings filled with a water-based liquid.) The pictures are sometimes almost transparent and other times blurry, but occasionally they are as natural and solid looking as the face of the living person that also appears in the picture. Macy has reproduced many of these pictures in his book *Spirit Faces*.

Research on ITC is spreading; the number of serious investigators is increasing. Father Brune, who has been surveying the field for many years, estimates that there may be as many as twenty thousand ITC researchers today in various parts of the world, most of them in the U.S. and Germany.

Communication beyond the grave is not limited to the instrumental voice form; there is also a noninstrumental, direct—that is, *telepathic—*

form. Often this form of after-death communication (ADC) occurs in dreams. Many people report that family members they lost recently appear to them in their sleep. Such dreams are generally dismissed as grief-induced hallucinations. However, the experience of anthropologist Marianne George should make us think again. George lived and worked among the Barok tribe in Papua New Guinea for nearly two years, between 1979 and 1985. Early during her first stay she was visited by the elderly shaman ("big woman") of the village. A short while later George had a dream in which the shaman spoke to her. In the morning, before the anthropologist saw anyone or had even left her hut, the sons of the shaman came to ask if she had understood what their mother had been saying. Later George had further dreams in which the shaman provided information she was seeking. The correctness of the information was borne out by subsequent inquiry, and the sons of the shaman always knew of the messages. (This kind of transpersonal dreaming is well known to the Barok—it is called *griman*.) Then, during the time George stayed with the tribe, the shaman died. Yet she continued to appear in George's dreams and continued to provide information. Also, the sons knew that their dead mother continued to appear in George's dreams.

After-death communication can also be induced with modern techniques. By using a simple technique, such as a series of rapid eye movements known as "sensory desensitization and reprocessing," psychiatrist Allan Botkin of the Center for Grief and Traumatic Loss in Libertyville, Illinois, has induced ADC in nearly three thousand patients. ADCs are obtained by about 98 percent of the people who try the experiment. Contact usually comes about rapidly, almost always in a single session. It is not limited or altered by the relationship of the experiencing subjects to the deceased. It also does not matter whether the subjects are deeply religious, agnostic, or convinced atheists. For the most part their experiences are clear, vivid, and convincing. The subjects find that their telepathic reconnection is real and shift nearly instantly from a state of grief to one of elation.

Telepathic communication with a deceased person is not unrelated to experiencing one's own consciousness while one's body is clinically dead: both suggest the persistence of consciousness beyond the brain. Investigation of so-called near-death experiences (NDEs) began with the groundbreaking work of Elisabeth Kübler-Ross in the 1960s. Since then an impressive number of cases of conscious experiences have been reported by patients who, for a brief period, were clinically dead. In studies carried out in major hospitals in Holland and England, nearly one fourth of the patients who suffered serious but not fatal heart attacks gave evidence of having had some form of consciousness with distinct perceptions during the time their EEG was flat. Physicians Pim van Lommel, Peter Fenwick, Sam Farnia, and B. Greyson have shown that the experiences recounted by the patients often correspond to what actually happened during the time they had zero brain function. Researcher Kenneth Ring found that consciousness at the portals of death gives rise to visual experiences that are essentially the same whether they are had by sighted people or by those who have been blind from birth.

The NDE phenomenon shows that some form of consciousness can persist even in the absence of a functioning brain. The frequent occurrences of OBEs (out-of-body experiences) also show that consciousness can be separated from the brain, at least for a time. Could consciousness also persist in people who have become entirely and irreversibly brain-dead—that is, in the deceased? Mediums such as James Van Praagh, John Edward, and George Anderson have mediated contact with thousands of deceased people and described the information they received from them; Raymond Moody collected a wide variety of "visionary encounters with departed loved ones." The possibility of authentic telepathic transcommunication, the same as authentic instrumental transcommunication, must be definitely admitted.

WHAT PREVENTS ITC'S RECOGNITION?

ITC's human, spiritual, and scientific implications are staggering. If it is authentic, why do we not hear more about it? Why does the press not report more on it, and why is it not a priority of scientific research?

A veil of ignorance surrounds the transcommunication phenomenon in the modern world. This has a number of reasons.

First, communication in the extrasensory or nonsensory mode can, but mostly does not, occur spontaneously, "out of the blue" (although the exceptions to this are so shattering that they change people's views of themselves and reality for the rest of their lives). In most cases nonsensory communication needs to be enabled—or at least facilitated—by entering into altered states of consciousness. These states are known; their physical correlates are EEG waves in the deeper frequency regions: alpha, theta, and occasionally even delta. Altered states occur spontaneously in daydreaming and in meditative, aesthetic, and religious experience, but the average person seldom makes use of them to experience nonsensory communication.

Altered states can also be induced. Psychiatrists, hypnotists, gurus, and spiritual leaders, the same as shamans and medicine men, use sensory reprocessing, hypnosis, breathing exercises, dancing, drumming, and in some cases psychedelic substances to induce them. Psychiatrist Stanislav Grof has induced altered states in thousands of patients, and the experiences reported by them are astounding.

Second, for ITC to occur, the brain/mind of the receiver may have to have a particular kind of sensitivity. This often calls for special abilities; as we have seen, known sensitives figure prominently among those who experience instrumental and other forms of transcommunication. The physical presence (if not the conscious awareness) of a sensitive is often a precondition for the occurrence of the phenomenon.

It is probable that in traditional societies more people possessed the required sensitivities. As anthropological research testifies, in traditional and non-Western cultures many experiences we consider "paranormal" are considered "normal"—for example, telepathy among Australian aborigines, past-life impressions in India, and communication with the spirits of ancestors by shamans in Africa, Siberia, and Latin America. The sensitivities in question may have been filtered out by Western civilization's reductionist and materialistic mindset.

Third, even though the ability to spontaneously receive nonsensory signals appears to have been largely lost, the number of occurrences of transcommunication in modern societies may still be considerably higher than those that are reported. This is very likely due to a bias inherent in the culture of modern society. In the Western world most of us tend to ignore, and even to repress from consciousness, phenomena that do not fit the belief encapsulated in the tenet of classical empiricism, "There is nothing in the mind that was not first in the eye." This is a powerful belief, and it is likely to limit the occurrence of ITC to people who either have a natural disposition to receiving information of nonsensory origin or are able to open their minds sufficiently not to repress the intuitions they may occasionally receive.

The myth of sense-reductionist empiricism affects most segments of Western society. In his thoroughly researched book *Is There An Afterlife?* David Fontana notes that resistance to questions relating to an afterlife spreads across four segments of society: established science, academic parapsychology, established religion, and the general public. Scientists, when questioned, often answer that they do not find the evidence for the survival of any form of spirit or consciousness convincing, yet for the most part they do not even know that the evidence exists—when confronted with it, they tend to dismiss it as mere superstition. Parapsychology, the very discipline created to investigate anomalous phenomena, often (but with notable exceptions) directs attention away from after-death phenomena: many parapsychologists fear that research into mediumship, apparitions, and other survival-related occurrences would hinder the acceptance of their discipline as a branch of science—a quest to which parapsychology has been painstakingly dedicated for three-quarters of a century. Established religion, in turn, typically equates communication from the beyond with witchcraft and the powers of evil and gives little if any factual information about what can be known or researched about the survival of spirit, soul, or consciousness, even though it proclaims this as a matter of dogma and faith.

The general public, however, doesn't so much oppose research into survival as avoids it: the entire subject has become taboo. In his book *The Ultimate Journey*, Stanislav Grof highlights this curious lack of interest in the phenomenon of death and all things connected with it. The cultural bias inherent in modern society militates not only against the study and recognition of after-death phenomena but also against all experiences related to death and dying. Modern people strongly believe that consciousness is an epiphenomenon of matter, produced by the brain. This concept profoundly influences beliefs, and even perceptions, in today's civilization. If consciousness cannot possibly exist independently of the living brain, communication beyond the grave must be pure fantasy.

EXPLORING AN EXPLANATION

No doubt, from a deeper spiritual and human point of view, the ITC phenomenon is extraordinarily important; it merits serious investigation in regard to the feasibility of finding a scientifically acceptable explanation. This requires connecting the phenomenon of transcommunication with an independently developed theory of the nature of consciousness and the possibility of its persistence following the deactivation of the brain with which it was associated. The theory of the information-recording, -conserving, and -transmitting cosmic plenum is a promising candidate for this research. The following is a first attempt to explain instrumental transcommunication as a natural process mediated by the Akashic field (A-field) of the cosmic plenum.

In connecting the ITC phenomenon with the theory of the A-field, the key principle is the transfer of information between holograms—in this case, between holograms generated by interfering wave fronts in the cosmic plenum. We can reconstruct how this information transfer may actually take place.

Scientists know that all objects emit waves at specific frequencies that radiate outward from the objects. When the wave field emanating from one object encounters another object, a part of it is reflected from that object and a part is absorbed by it. The object becomes energized and creates another wave field that moves back toward the object that emitted the initial wave field. The interference of the initial and the response wave fields creates an overall pattern, and this pattern carries information on the objects that created the fields. The interference pattern is effectively a hologram. The information carried in it is available at all points where the constitutive wave fields penetrate. It can be transferred from hologram to hologram if they resonate at the same frequency or at compatible frequencies.

Things in space and time are embedded in the electromagnetic field; the waves they emit are EM waves. However, things in space and time are also embedded in the cosmic plenum, and in that deeper dimension they create waves of a different kind: most likely scalar waves (these, as noted in part 2, are nonvectorial waves of pure magnitude; they carry information but not energy). The interference patterns of these waves form holograms that endure indefinitely in the plenum. The information they carry remains available for exchange with holograms resonating at compatible frequencies.

These considerations offer a basis for elaborating a scientific explanation of instrumental transcommunication. In regard to the experience recounted here, the explanation is that Bacci's brain and nervous system are able to enter a frequency domain compatible with holographically stored information in the A-field of the plenum. For the past four decades Bacci has made use of this ability and has been receiving paranormal messages.

Unlike most mediums, mystics, and other intuitive people, Bacci receives the messages not telepathically, in an altered state of consciousness, but in his usual state of awareness in physical contact with an old-fashioned radio. As he touches the radio's wooden frame, the receiver becomes tuned to the frequencies of his brain and nervous system. When in his search of the shortwave broadcast band he comes across the appropriate frequency, the radio transmits signals from plenum-based holograms rather than from shortwave broadcast stations. The holograms accessed by Bacci carry information corresponding to the consciousness of recently deceased persons.

This general theoretical framework needs to be filled in with answers to several questions. We start with some specific, comparatively technical queries.

1. How does a discarnate consciousness access information from the world of the living?

Given that the consciousness implied by the phenomenon of transcommunication is no longer associated with a living body, it does not have access to sensory organs. How, then, does it access information from living people?

We should note that the evidence for transcommunication does not suggest that a discarnate consciousness has sensory types of perceptions, such as seeing sights in three dimensions and hearing ordinary sounds associated with the sights. It does indicate, however, that such a consciousness can perceive questions and comments from living persons. An answer to this puzzle requires recourse to the theory of holographic information exchange outlined above. Holograms tuned to the same frequency can exchange information. This tells us that transcommunication occurs when the hologram that carries the lifetime experiences of a now defunct person enters into adaptive resonance with the hologram created by the brain of the living interlocutor. Under such circumstances the deceased person's A-field hologram accesses some elements of the information encoded in the A-field by the brain of the living interlocutor.

2. How does a discarnate consciousness produce audible sounds in a radio?

The concept of information exchange among holograms resonating at the same frequency offers a plausible answer to the above question as well. It is not that the radio would receive the signals emitted by a plenum-based hologram, for that assumption is unsupported by both theory and experience. It would presuppose that information in the cosmic plenum can produce direct physical effects, such as creating signals in the electromagnetic field. There is no independent evidence that this is the case.

In the case of Marcello Bacci, the assumption that it is the radio that receives the signals is counterindicated by three experimentally verified facts:

1. The radio fails to produce the voices in Bacci's absence.
2. The production of the voices—unlike the normal operation of the radio—does not involve an increase in the ambient electrical and magnetic field.
3. The voices continue without alteration even when the frequency modulation and the local oscillation valves are removed from the radio.

The fact that the radio produces the anomalous voices only in Bacci's presence matches observations in other ITC experiments. In the cases mentioned above—the transcommunication produced by Konstantin Raudive, Otto König, and Anabela Cardoso (as well as in a number of other instances, inter alia the experiments of Friedrich Jürgenson, Raymond Bayless, and Attila von Szalay)—the results were obtained only in the presence of psychically gifted experimenters. In some well documented cases—for example, that of Peter Härting of the Darmstadt group, who had long-lasting and thoroughly documented contact with the discarnate consciousness known as ABX JUNO, and that of William O'Neil of the Metascience Foundation, who had enduring communication with the deceased George Jeffries Mueller—regular transcommunication ceased abruptly when the living communicator died.

These observations support the hypothesis that contact between the living partner and a discarnate consciousness requires special "tuning" or harmonization. In many cases the living partner has privileged contact with one specific "spirit communicator" even if he or she picks up voices (or images) from other discarnate consciousnesses as well. (For example, Styhe was the standard communicator for S. W. Estep, Hyppolite Baraduc for Vladimir Delavre, ABX JUNO for Peter Härting, and the "Technician" for Swejen Salter. For Bacci the privileged communicator has been Cordula, although in the experiment reported here she did not manifest herself.) When this fine-tuned relationship ceases, the spirit communicator can sometimes find another living partner: in the case of George Jeffries Mueller as well as in that of Konstantin Raudive (who had earlier communicated mainly with Jules and Maggy Harsch-Fischbach of the

Luxembourg group) this turned out to be Adolf Homes of Rivenich, Germany. (It is worthy of note that in his communication with Mueller, Homes received the following message: "The dialogue ceases with the death of the researcher because the vibration required for this is no longer given.")

It appears that the living partner—usually a psychically gifted individual—needs to be physically present, but not necessarily consciously aware that transcommunication is taking place. This was shown by a famous case described by the Brazilian investigator Oscar d'Argonnel. He had been receiving voice messages on the telephone and went to some trouble to ascertain that they were of paranormal origin. In January 1919 he was visiting Abelardo, a psychic, who was having lunch. D'Argonnel wanted to call his friend Figner on the telephone and did so in the room next to where Abelardo was having his meal. While d'Argonnel was talking to Figner on the phone, strange noises began to come through, and the voice of Father Manoel, who had died some years before then, came on the line. D'Argonnel and Figner had a long and clear three-way conversation with Manoel during which d'Argonnel looked periodically at Abelardo to see whether he was following it, but the medium was engrossed in his meal. D'Argonnel asked Father Manoel on the phone if he should call Abelardo, but the latter replied, "No, for I cannot stay." Manoel stayed on the line for a while longer and then took formal leave: "Adieu, d'Argonnel, Adieu, Figner." D'Argonnel finished the conversation with Figner and then went to Abelardo to tell him what had transpired. Abelardo did not know anything about it. It appears that the presence of the psychic Abelardo in the next room was sufficient to produce the ITC phenomenon—it did not require his conscious attention.

Because it is possible that the living partners in ITC are not consciously aware that transcommunication is taking place, skeptical investigators have suggested that it is their subconscious mind that produces the voices (or the images). This, however, is counterindicated by the observation that the messages often contain information to which the living partners had no access—for example, descriptions of distant objects and places and languages unknown to them.

The so-called animist hypothesis—that the subconscious mind of the medium is responsible for the ITC phenomenon—goes part of the way toward explaining what takes place, but it does not go far enough. The consciousness (or the subconscious mind) of the investigator is indeed involved in the communication and appears to be essential for it, but it does not *produce* the voices (or images); it merely *transmits* them. The phenomena originate as holographic information in the A-field of the cosmic plenum. The brain and nervous system of the experimenters pick up this information and transmit it to the instrument in a form the latter can convert into sound or image. This is not unprecedented. In an impressive number of cases psychic individuals have proven able to project some form of information from their conscious or subconscious mind: voices to tape recorders, images to TV screens. Even if the physics underlying the process is not elucidated, the process itself appears authentic.

An explanation in terms of physics should, however, be feasible. A useful consideration for attempting it is that a radio tuned to empty regions of the shortwave band (or operating without the frequency modulator and local oscillation valves)—like a TV set tuned to empty regions of the broadcast band—is a system in a state of chaos. It produces random static. In this condition it is ultrasensitive, and it is conceivable that impulses arriving from the human brain and nervous system (transmitted in Bacci's case by hands-on contact with the radio) are physically of a kind that can be transformed into sound waves by an electronic instrument at the edge of chaos.

We now turn to the truly fundamental question.

3. How does a consciousness that is no longer associated with a living brain persist in the cosmic plenum?

This question still lacks a scientifically acceptable answer, but approaches to its investigation can be indicated already. We explore a promising approach through the theory of the A-field.

A-field theory, we have seen, tells us that the holographic traces of the consciousness associated with a living brain are conserved in the cosmic plenum. This is a beginning, but we need to go further. Transcommunication suggests that the plenum contains not merely a passive record of a person's consciousness, created during that person's lifetime and then persisting unchanged, but harbors a dynamic bundle of information based on the experiences accumulated in that lifetime. Under suitable conditions this bundle of information is capable of autonomous development despite the demise of the body.

How is this possible? Various hypotheses of this phenomenon have been advanced by mystics as well as ITC researchers. Among the most widely discussed hypotheses, the concept of multiple "shells" surrounding the body—physical, mental, and spiritual—merits special attention. According to esoteric wisdom, several shells compose the human being, one embedded in the other, like the skins of an onion. When an individual dies it is not only his mental shell (his consciousness) that leaves his physical shell (his body); his spiritual shells separate from it as well. These shells separate gradually, in stages. In the first stages some of the persisting spiritual shells, or shell fragments, still carry the thoughts, feelings, desires, and memories of the deceased. They remain active and possess a degree of autonomy even when separated. When contacted they produce responses on their own.

Another hypothesis links the phenomenon of transcommunication with modern physics. It views discarnate consciousnesses as waves that are not perceptible by the sensory organs but are nonetheless real. Supporters of this hypothesis point out that many of the waves that propagate in space (quantum waves, gravity waves, and others) cannot be directly perceived but must be deduced through complex chains of reasoning. Some waves—such as EM waves within a specific range—can be transformed into sense-perceivable form by electronic instruments. There is no reason to assume that there may not be waves that we have not yet discovered for lack of the necessary instruments or theoretical frameworks. Some of these waves could be accessed by specially gifted "psychic" individuals. When accessed they would produce telepathic (direct) or instrumental (indirect) transcommunication.

The two hypotheses mentioned above—that of shells or shell fragments that leave the body in stages and of imperceptible but real waves—can also be combined. The shells—for example, our spiritual shell, also known as our "etheric body"—could enter a larger wave field and integrate with other shells in that field. The renowned mystic Alice Bailey suggested something along these lines. She wrote, "This word 'ether' is a generic term covering the ocean of energies which are all interrelated and which constitute that one synthetic energy body of our planet . . . the etheric or energy body, therefore, of every human being is an integral part of the etheric body of the planet itself."

Gustav Fechner, the pragmatic founder of experimental methods in psychology, advanced an analogous hypothesis. "When one of us dies," he wrote after recovering from a serious illness, "it is as if an eye of the world were closed, for all perceptive contributions from that particular quarter cease. But the memories and conceptual relations that have spun themselves round the perceptions of that person remain in the larger Earth-life as distinct as ever, and form new relations and grow and develop throughout all the future, in the same way in which our own distinct objects of thought, once stored in memory, form new relations and develop throughout our whole finite life."

The above hypotheses offer a promising approach toward a scientific explanation of transcommunication. This explanation would differ in detail from the esoteric tradition, according to which a discarnate consciousness possesses the degree of autonomy required for a dialogue with living persons because at death the spiritual shell or etheric body detaches itself from the grosser layers and conserves the consciousness of the individual, perhaps integrated with the larger consciousness that embeds the Earth. It would agree with the hypothesis that imperceptible waves are involved in transcommunication, but the answer it would suggest is simpler and at the same time more general. There is no need to assume a special spiritual shell in regard to human beings; all things in space and time, from quanta to galaxies, leave their traces in the plenum. These traces constitute natural holograms. In the plenum the holograms are not subject to attenuation or cancellation. As new wave fronts are generated, the existing holograms superpose and the information they contain is conserved; it is not overwritten but integrated in the manner of multiplex holograms.

Here we have the basis for a cogent explanation of the conservation of information in the cosmic plenum. The next step is to discover how the conserved information is capable of autonomous development. This step is yet to be accomplished; it is a difficult but not insoluble task. Given the theoretical tools, the mathematics, and the electronic simulation methods at our disposal, it should not be impossible to discover how complex sets of coherent elements within an information-rich extremely complex field can function with a form and level of autonomy that permits creating fresh information based on the information already given.

The logical place to seek a scientific explanation of the mystery of transcommunication is not the metaphysics of souls and spirits but the physics of complex field theory.

Bibliography

Adi Da. *Not-Two Is Peace*: *The Ordinary People's Way of Global Cooperative Order*. Middletown, Calif.: IS Peace 723, 2007.

Afshar, Shahriar. As reported in Marcus Chown, "Quantum rebel." *New Scientist*, no. 2457 (July 2004).

Akimov, A. E., and G. I. Shipov. "Torsion fields and their experimental manifestations." *Journal of New Energy* 2, no. 2 (1997).

Akimov, A. E., and V. Ya. Tarasenko. "Models of polarized states of the physical vacuum and torsion fields." *Soviet Physics Journal* 35, no. 3 (1992).

d'Argonnel, Oscar. *Vozes do alem pelo telephone*. Rio de Janeiro, 1925.

Bailey, Alice. *Telepathy and the Etheric Vehicle*. New York: Lucis Trust, 1950.

Botkin, Allan, and R. Craig Hogan. *Reconnections: The Induction of After-Death Communication in Clinical Practice*. Charlottesville, Va.: Hampton Roads, 2006.

Cade, C. Maxwell. *The Awakened Mind: Biofeedback and the Development of Higher States of Awareness*. New York: Delacorte Press, 1979.

Cardoso, Anabela. "Survival research." *Journal of Conscientology* 6, no. 21 (2003).

Clayton, Philip D. *God and Contemporary Science*. Grand Rapids, Mich.: Eerdmans, 1997.

Clifford, William. "On the Space Theory of Matter." In *The World of Mathematics*, 568. New York: Simon and Schuster, 1956. Cited by Milo Wolff and Geoff Haselhurst, "Einstein's Last Question." *VIA: Journal of Integral Thinking for Visionary Action* 3, no.1 (2005).

Delavre, Vladimir. "Paranormal transferphänomene." *Transkommunikation* 1, no. 4 (1992).

Durning, Alan. *How Much Is Enough?* New York: Norton, 1992.

Engel, Gregory S., T. S. Calhoun, Elizabeth Read, Tae-Kyu Ahn et al. "Evidence for Wavelike Energy Transfer through Quantum Coherence in Photosynthetic Systems." *Nature* 446 (12 April 2007).

Einstein, Albert. "The Concept of Space." *Nature* 125 (1930).

Fechner, Gustav. Quoted in William James, *The Pluralistic Universe*. London: Longmans, Green & Co., 1909.

Festa, Mario. "A Particular Experiment at the Psychophonic Centre in Grosseto directed by Marcello Bacci." *ITC Journal* 10 (2004): 27–31.

Fiscaletti, Davide, and Amrit Sorli. "A-Temporal Physical Space and Quantum Nonlocality." *Electronic Journal of Theoretical Physics* 2, no.6 (2005): 15–20. www.ejtp.com.

———. "Toward a New Interpretation of Subatomic Particles and their Motion inside A-Temporal Physical Space." *Frontier Perspectives* 15, no.2 (2006).

Fontana, David. *Is There An Afterlife? A Comprehensive Overview of the Evidence*. Ropley, Hampshire, England: O Books, 2006.

———. "Why the Opposition to Evidence for Survival?" *Network Review* 3 (Spring 2007).

Fox, Warwick. "A Critical Overview of Environmental Ethics." *World Futures* 46 (1996): 1–21.

Fuller, J. G. *The Ghost of 29 Megacycles*. London: Souvenir Press, 1985.

Gazdag, László. "Superfluid mediums, vacuum spaces." *Speculations in Science and Technology* 12, no.1 (1989).

George, Marianne. Quoted in Paul Devereux, "The Movable Feast," in *Mind Before Matter*, edited by T. Pfeiffer, J. Mack, and P. Devereux. Ropley, Hampshire, England: O-Books, 2008.

Greyson, B. "Incidence and correlates of near-death experiences in a cardiac care unit." *General Hospital Psychiatry* 25 (2003): 269–76.

Grof, Stanislav. *The Cosmic Game: Explorations at the Frontiers of Human Consciousness*. Albany: SUNY Press, 1999.

———. *Psychology of the Future*. Albany: SUNY Press, 2000.

———. *The Ultimate Journey: Consciousness and the Mystery of Death*. Ben Lomond, Calif.: Multidisciplinary Association of Psychedelic Studies, 2006.

Haisch, Bernhard, Alfonso Rueda, and H. E. Puthoff. "Inertia as a zero-point-field Lorentz force." *Physical Review A* 49.2 (1994).

Hauser, Marc. *Moral Minds*. New York: HarperCollins, 2006.

Ives, Herbert. "Extrapolation from the Michelson-Morley experiment." *Journal of the Optical Society of America* 40 (1950).

———. "Light signals sent around a closed path." *Journal of the Optical Society of America* 2 (1938).

———. "Lorentz-type transformations as derived from performable rod and clock operations." *Journal of the Optical Society of America* 39 (1949).

———. "Revisions of the Lorentz transformations." *Proceedings of the American Philosophical Society* 95 (1951).

Kubis, Pat, and Mark Macy. *Conversations Beyond the Light.* Boulder, Colo.: Griffin Publishing and Continuing Life Research, 1995.

Kübler-Ross, Elisabeth. *On Death and Dying.* New York: Macmillan, 1969.

Laszlo, Ervin. *The Choice: Evolution or Extinction.* New York: Tarcher/Putnam, 1994.

———. *The Connectivity Hypothesis.* Albany: SUNY Press, 2003.

———. *The Creative Cosmos.* Edinburgh: Floris Books, 1993.

———. *Evolution: The General Theory.* Cresskill, N.J.: Hampton Press, 1996.

———. *The Interconnected Universe.* River Edge, N.J.: World Scientific, 1995.

———. *Science and the Akashic Field,* 2nd ed. Rochester, Vt.: Inner Traditions, 2007.

———. *Science and the Reenchantment of the Cosmos.* Rochester, Vt.: Inner Traditions, 2006.

———. *The Systems View of the World.* Cresskill, N.J.: Hampton Press, 1996.

———. *The Whispering Pond.* Rockport, Mass.: Element Books, 1996.

Lommel, Pim van. "Near-death experience, consciousness, and the brain." *World Futures* 62, nos. 1–2 (2006).

Lommel, Pim van, R. Van Wees, V. Meyers, and I. Eifferich. "Near-death experience in survivors of cardiac arrest: a prospective study in the Netherlands." *Lancet* 358 (2001).

Macy, Mark. *Spirit Faces: Truth About the Afterlife.* Newburyport, Mass.: Weiser Books, 2006.

Marcer, Peter, and W. Schempp. "Model of the neuron working by quantum holography." *Informatica* 21 (1997).

Mitchell, Edgar R. *The Way of the Explorer: An Apollo Astronaut's Journey through the Material and Mystical Worlds.* New York: Putnam, 1996.

Moody, Raymond. *Visionary Encounters with Departed Loved Ones.* New York: Ballantine Books, 1994.

Parliament of the World's Religions. *Towards a Global Ethic.* Council of the Parliament of the World's Religions (August 28–September 5, 1993).

Parnia, Sam, and Peter Fenwick. "Near-death experiences in cardiac arrest: visions of a dying brain or visions of a new science of consciousness." *Resuscitation* 52 (2002): 5–11.

Persinger, Michael, and Stanley Krippner. "Dream ESP Experiments and Geomagnetic Activity." *The Journal of the American Society for Psychical Research* 83 (1989).

Puthoff, Harold. "Gravity as a zero-point-fluctuation force." *Physical Review A* 39 (1989): 2333; *Physical Review A* 47 (1993): 3454.

———. "Ground state of hydrogen as a zero-point-fluctuation-determined state." *Physical Review D* 35, no. 10 (1987).

———. "Source of vacuum electromagnetic zero-point energy." *Physical Review A* 40, no. 9 (1989).

Raudive, Konstantin. *Breakthrough.* Gerrards Cross, England: Colin Smythe, 1971.

Ring, Kenneth. "Near-death and out-of-body experiences in the blind: A study of apparent eyeless vision." *Journal of Near-Death Studies* 16 (Winter 1997).

Rueda, Alfonso, and Bernhard Haisch. "Inertia as reaction of the vacuum to accelerated motion." *Physics Letters A* 240 (1998).

Sagnac, G. "The luminiferous ether demonstrated by the effect of the relative motion of the ether in an interferometer in uniform rotation." *Comptes Rendus de l'Académie des Sciences* 157 (1913).

Schäfer, Lothar. "Quantum Reality, the Emergence of Complex Order from Virtual States, and the Importance of Consciousness in the Universe." *Zygon* 41 (September 2006).

Schrödinger, Erwin. *Schrödinger: Life and Thought.* London: Cambridge University Press, 1989.

Shipov, G. I. *A Theory of the Physical Vacuum: A New Paradigm.* Moscow: International Institute for Theoretical and Applied Physics RANS, 1998.

Silvertooth, Ernest W. "Experimental detection of the ether." *Speculations in Science and Technology* 10 (1987).

———. "Motion through the ether." *Electronics and Wireless World* (May 1989).

————. "A new Michelson-Morley experiment." *Physics Essays* 5 (1992).

Smolins, Lee. *The Trouble With Physics: The Rise of String Theory, the Fall of a Science, and What Comes Next.* New York: Houghton Mifflin, 2006.

Strijbos, S. "Ethics for an Age of Social Transformation." *World Futures* 46, no. 3 (1996).

The Union of Concerned Scientists. *World Scientists' Warning to Humanity.* April 1993.

Waal, Frans de. *Primates and Philosophers.* Princeton: Princeton University Press, 2006.

Weiskel, Timothy C. "Environmental ethics: conserving views on a small planet." The Winter Colloquium Series, The Morrison Institute for Population and Resource Studies, 31 January 1990.

Wheeler, John A. In *The Ghost in the Atom*, edited by P. C. W. Davies and J. R. Brown. Cambridge: Cambridge University Press, 2000.

Whitehead, Alfred North. *Process and Reality: An Essay in Cosmology.* New York: Macmillan, 1929; corrected edition, 1978.

Wolff, Milo. "Cosmology, the Quantum Universe, and Electron Spin." In *Gravitation and Cosmology: From the Hubble Radius to the Planck Scale*, edited by R. I. Amoroso et al. Amsterdam: Kluwer Academic, 2002.

Index